AQUACULTURE - PLANTS AND INVERTEBRATES

Edited by **Genaro Diarte-Plata** and **Ruth Escamilla-Montes**

Aquaculture - Plants and Invertebrates
http://dx.doi.org/10.5772/intechopen.74183
Edited by Genaro Diarte-Plata and Ruth Escamilla-Montes

Contributors

Ramachandra Reddy Pamuru, Kishori Battina, Gulnihal Ozbay, Stacy Smith, Ibrahim Adeshina, José Manuel Mazón-Suástegui, Dariel Tovar-Ramirez, Joan Sebastian Salas Leiva, Jesús Antonio López Carvallo, Milagro Garcia-Bernal, Fabiola Arcos, Carmen Rodríguez-Jaramillo, Pablo Ormart, Laura Elizabeth Ibarra García Ibarra García, María Araceli Avilés-Quevedo, Andressa Teles, José Luis García Corona, Ulysses Barajas Ponce, Reinaldo González González, Pedro Saucedo Lastra, Antonia Del Pilar Rosero-García, Fernando Abasolo, José Delfino Barajas Frías, Angel Campa-Córdova, Nadia Libia Ortíz Cornejo, Jorge Luis Tordecillas Guillén, Francisco Antonio Alvarez-Gil, Gustavo Pineda-Mahr, Jesús Peiro-López, Miguel Robles-Mungaray Miguel, Genaro Diarte-Plata, Ruth Escamilla-Montes

First published in London, United Kingdom, 2019 by IntechOpen
IntechOpen is the global imprint of INTECHOPEN LIMITED, registered in England and Wales, registration number: 11086078, The Shard, 25th floor, 32 London Bridge Street
London, SE19SG – United Kingdom
Printed in Croatia

British Library Cataloguing-in-Publication Data
A catalogue record for this book is available from the British Library

Additional hard copies can be obtained from orders@intechopen.com

Aquaculture - Plants and Invertebrates, Edited by Genaro Diarte-Plata and Ruth Escamilla-Montes
p. cm.
Print ISBN 978-1-78984-996-7
Online ISBN 978-1-78984-997-4

For EU product safety concerns:
IN TECH d.o.o., Prolaz Marije Krucifikse Kozulić 3, 51000 Rijeka, Croatia,
info@intechopen.com or visit our website at intechopen.com.

Meet the editors

Dr. Genaro Diarte-Plata is a candidate of SNI-CONACYT Mexico. He is a biologist in Aquaculture at the Universidad Autónoma de Sinaloa, Mexico. He has a Master degree in Natural Resources and Environment from the Instituto Politécnico Nacional (IPN CIIDIR-Sinaloa), Mexico. He is a Professor at the Instituto Politécnico Nacional, Mexico (2006 to present). He is a researcher of the Sinaloa System of Researchers and Technologists (SSIT-INAPI Sinaloa). He received his Doctorate in Marine and Coastal Sciences at the Universidad Autónoma de Baja California Sur, Mexico. His interests are in investigations in mariculture, culture and ecology of molluscs and crustaceans. As Professor-Researcher within the IPN, he has participated in 45 research projects, given 10 seminars, 250 presentations at international and national conferences, written 5 book chapters and 17 indexed articles. He teaches Descriptive and Multivariate Statistics at the Bachelor level, and Aquaculture at the Postgraduate level. He is the director of 10 undergraduate theses and 3 Masters and 1 Doctorate from various universities.

Dr. Ruth Escamilla-Montes is a Doctor in Marine and Coastal Sciences with a specialty in Aquaculture. She is the Assistant Technologist of the Sinaloa System of Researchers and Technologists. Her areas of study are cultivation and ecology of molluscs and crustaceans, and in the last five years has focused on the isolation and characterization of lactic acid bacteria, yeasts and bacilli with probiotic potential and pathogenic bacteria such as *Vibrio parahaemolyticus* and its effect on the growth, survival and expression of genes of the immune system and stress of marine organisms. She has participated in 8 research projects, as well as presentations at national and international conferences, 4 book chapters and 23 articles in ISI Thomson-JCR international journals. She was an arbitrator in the revision of manuscripts sent to International Journals (ISI-JCR).

Contents

Preface

Aquaculture, in recent years, has gained importance from the point of view of food production. Nowadays, it is a dynamic, profitable and constantly developing field that covers a wide variety of production systems, carried out on a great diversity of organisms. It also faces important challenges related to the characteristics of the aquatic environment and the use of its resources, but also presents unique opportunities to consolidate as an efficient way to supply increasing quantities of protein and other high quality nutrients to a human population in constant growth.

This book is made up of three sections. Following the "Introduction", the second section on "Sustainable Aquaculture" offers information related to integrated activity of rice cultivation and aquaculture practices that bases studies on the production of grains (carbohydrates) and animal protein without affecting the quality and quantity of grain yield in the plots, and hence resulting in extra profits for the producers, with these activities being more popular in Asian countries than in Western countries. In addition, the participation of aquaculture in the restoration of the fishery numbers of an important species is evaluated, as is the American oyster *Crassostrea virginica*. The third section "Homeopathy and Probiotics" presents an alternative medical system proposed in the XVIII, which uses highly diluted and agitated substances derived from plants, minerals or animals, which have proven their effectiveness in human medicine, agronomy, veterinary, and are currently new in the area of marine aquaculture. In addition, we present a study focused on the growth performance and nutrient utilization of the freshwater shrimp *Macrobrachium vollenhovenii* fed diets supplemented with *Lactobacillus acidophilus*. This book can be used by students, professors and researchers in the area biological sciences.

Dr. Genaro Diarte-Plata and Dr. Ruth Escamilla-Montes
Instituto Politécnico Nacional
Centro Interdisciplinario de Investigación para el Desarrollo Integral Regional
Sinaloa, México

Section 1

Introduction

Chapter 1

Introductory Chapter: Importance of Plant and Invertebrates in Aquaculture

Ruth Escamilla-Montes and Genaro Diarte-Plata

Additional information is available at the end of the chapter

http://dx.doi.org/10.5772/intechopen.81762

1. Introduction

Aquaculture is an economic activity that presents a rhythm of global production with a sustained growth of 10–11% per year during the last years, with expectations of equating production by extractive fishing in 2025 [1]. It currently contributes about 50% of the world fish supply, and is considered to be one of the main economic activities of this century. In 2010, world aquaculture production was around 59.9 million tons; where freshwater fish dominated production (56.4%), followed by the cultivation of invertebrates such as mollusks (23.6%) and crustaceans (9.6%) [2].

It is an activity that encompasses very varied practices, and a wide range of species, systems and production techniques. Aquaculture can be defined as the production of aquatic organisms with techniques aimed at making their performance more efficient. It is worth mentioning that more than half of the total amount of food of aquatic origin consumed today by the world population, as well as products destined for non-food uses, comes from aquaculture farms where fish, crustaceans, micro algae, mollusks and other invertebrates are raised [1, 4, 5]. Particularly, invertebrates make up 95% of the animals that inhabit our planet. Due to its great biodiversity, it has not been possible to study in its entirety. There are species of mollusks, crustaceans and echinoderms that are consumed by man and therefore are the object of artisanal or massive fishing, which can lead to problems of population reductions, local extinctions or loss of genetic diversity [3], thus Aquaculture is a good alternative to solve this problem.

2. Sections of the book

The capture and cultivation of aquatic organisms in paddy fields has a long history and tradition, especially in Asia, where the availability of rice and fish has been linked to prosperity and food security. Rice-based ecosystems provide habitats for a wide variety of aquatic organisms used extensively by the local population. They also allow the improvement and breeding of aquatic organisms. A wide variety of aquatic species such as carp, tilapia, catfish and breams are being raised in the rice fields. Prices and market preferences can provide decisive opportunities for farmers to further diversify the use of species, especially eels, loaches and various crustaceans, to sell and market higher value biological products [7]. The problems associated with the breeding of aquatic organisms in paddies do not differ from those related to the development of aquaculture in general. These include the availability and access to seeds, feed and capital, as well as natural risks related to water control, diseases and predation [8].

Molluscs are currently the group of cultivable marine organisms that offers better prospects in terms of production and economic profitability, their production costs are not high and are a valuable source of food. To perform its cultivation requires detailed knowledge of the basic biology of the species, supply sources of seed, growth parameters and mortality in culture, and the effect of environmental conditions as well as their spatial and temporal variability. The cultivation of bivalve mollusks represents an economically viable alternative due to the possibility of large-scale operation, in addition, this activity can be environmentally sustainable by helping to reduce fishing effort in coastal areas [6].

The constant increase in world population necessarily implies a challenge in terms of food production in large volumes and with high nutritional quality. To achieve the economically profitable production of healthy animals with a limited environmental impact, it is necessary to improve growth rates and feeding and reproduction efficiency, decreasing the losses caused by diseases, by improving the immune response, diagnostic techniques and prophylactic measures [6, 9]. In an intensive aquatic production system, control of the disease plays a key role, where an intimate relationship between the host and bacteria is present [10, 11]. There is currently a widespread concern that antibacterial agents in aquaculture will lead to the emergence of antibiotics resistant bacteria [12]. Probiotics and the use of homeopathy, which act mainly on the innate response of cultured organisms, constitute a viable, promising and economic strategy to make aquaculture process more sustainable, since it reduces the indiscriminate use of antibiotics and chemotherapeutic product.

In case of treatment with probiotic, it has been carried out successfully in mollusks [13], fish [14, 15] and crustacean species [11, 16, 17]. Wherein the probiotics used in aquaculture studies include Gram-positive and Gram-negative bacteria, bacteriophages, yeasts and unicellular algae [18], and the beneficial effects include growth and feeding efficiencies in culture systems [19].

Homeopathy in the aquaculture of freshwater and marine species is a potential alternative for the world aquaculture industry, because their medicines are free of relevant adverse reactions and do not bio-accumulate toxic substances in the harvested product. The studies realized, although scarce, suggest that homeopathy can be applied with prophylactic and therapeutic criteria. Among its mechanisms of probable action stands out the stimulation of the innate and acquired immune system, and consequently the increase in the resistance of the treated organism, against the pathogens that normally proliferate proportionally to the level of intensification of the culture. Higher survival, growth and reduction of stress levels have been reported, as well as notable changes in other parameters observed, such as less inclusion of lipids in the liver, greater hypertrophy of the muscle fiber, production of mucin-producing cells that are related to the inhibition to the entrance of parasites, and changes in blood parameters. These are indicators of improvement in health and nutrition of the organism cultivated, and if all this can be achieved by applying "ultra-diluted" doses, production costs are reduced and harmful effects are mitigated to the environment making aquaculture homeopathy an eco-sustainable alternative [20].

Conflict of interest

We declare no conflict of interest.

Author details

Ruth Escamilla-Montes and Genaro Diarte-Plata*

*Address all correspondence to: gdiarte@ipn.mx

Instituto Politécnico Nacional, CIIDIR Sinaloa, Mexico

References

[1] Tacon AJ. Aquaculture production trends analysis. In: Review of the State of World Aquaculture. Roma, Italy: Food and Agriculture Organization of the United Nations (FAO Fisheries Circular No. 886); 2003. pp. 5-29

[2] FAO Fisheries Department. State of world fisheries and aquaculture 2012. In: FAO Fisheries Technical Paper. Rome: FAO; 2012

[3] Bigatti G, Penchaszadeh PE. Invertebrados del mar patagónico, Diagnóstico de la problemática actual y potencial de su conservación y manejo. In: Estado de Conservación del Mar Patagónico. Foro para la conservación del Mar Patagónico y Áreas de Influencias. Referencias; 2008. pp. 105-130

[4] Muller-Feuga A. Microalgae for aquaculture: The current global situation and future trends. In: Richmond A, editor. Handbook of Microalgal Culture. Oxford: Blackwell Science; 2004. pp. 352-364

[5] Freites L, Lodeiros C, Guevara M, Alió J, Graziani C. Experiencias en el cultivo de organismos marinos en el Golfo de Cariaco, Venezuela saber. Revista Multidisciplinaria del Consejo de Investigación de la Universidad de Oriente. 2012;**24**:5-24

[6] Pipitone C, Badalamenti F, Anna GD, Patti B. Fish biomass increase after a four-year trawl ban inthe Gulf of Castellammare (NW Sicily, Mediterranean Sea). Fisheries Research. 2000;**48**:23-30

[7] Miao WM. Recent developments in rice-fish culture in China: A holistic approach for livelihood improvement in rural areas. In: de Silva SS, Davy FB, editors. Success Stories in Asian Aquaculture. Springer; 2010. pp. 15-42

[8] Lu J, Li X. Review of rice–fish-farming systems in China—One of the globally important ingenious agricultural heritage systems (GIAHS). Aquaculture. 2006;**260**(1-4):106-113

[9] Akira S, Uematsu S, Takeuchi O. Pathogen recognition and innate immunity. Cell. 2006; **124**:783-801

[10] Carnevali O, Zampon MCI, Sulpizio R, Rollo A, Nardi M, Orpianesi C, et al. Administration of probiotic strain to improve sea bream wellness during development. Aquaculture International. 2004;**12**:377-386

[11] Rodríguez J, Espinosa Y, Echeverría F, Cárdenas G, Román R, Stern S. Exposure to probiotics and β-1,3/1,6-glucans in larviculture modifies the immune response of *Penaeus vannamei* juveniles and both the survival to white spot syndrome virus challenge and pond culture. Aquaculture. 2007;**273**:405-415

[12] Scholz U, García-Díaz G, Ricque D, Cruz-Suárez LE, Vargas-Albores F, Latchford J. Enhancement of vibriosis resistance in juvenile *Penaeus vannamei* by supplementation of d resistance in juvenile *Penaeus vannamei* by supplementation of diets with different yeast products. Aquaculture. 1999;**176**:271-283

[13] Macey BM, Coyne VE. Colonization of the gastrointestinal tract of the farmed South African abalone *Haliotis midae* by the probionts *Vibrio midae* SY9, *Cryptococcus* sp. SS1, and *Debaryomyces hansenii* AY1. Marine Biotechnology. 2006;**8**:246-259

[14] Robertson PAW, O'Dowd C, Burrels C, Williams P, Austin B. Use of *Carnobacterium* sp. as a probiotic for Atlantic salmon (*Salmo salar* L.) and rainbow trout (*Oncorhynchus mykiss,* Walbaum). Aquaculture. 2000;**185**:235-243

[15] Brunt J, Newaj-Fyzul A, Austin B. The development of probiotics for the control of multiple bacterial diseases of rainbow trout, *Oncorhynchus mykiss* (Walbaum). Journal of Fish Diseases. 2007;**30**:573-579

[16] Harzevili ARS, Van Duffel H, Dhert P, Swings J, Sorgeloos P. Use of a potential probiotic *Lactobacillus lactis* Ar21 strain for the enhancement of growth in the rotifer *Brachionus plicatilis* (Müller). Aquaculture Research. 1998;**29**:411-417

[17] Rengpipa TS, Rukpratanporn S, Piyatiratitivorakul S, Menasav P. Immunity enhancement in black tiger shrimp (*Penaeus monodon*) by a probiont bacterium (*Bacillus* S11). Aquaculture. 2000;**191**:271-288

[18] Irianto A, Austin B. Probiotics in aquaculture. Journal of Fish Diseases. 2002;**25**:633-642

[19] Venkat HK, Shau NP, Jain KJ. Effect on feeding Lactobacillus-based probiotics on the gut microflora, growth and survival of postlarvae of *Macrobrachium rosenbergii* (de Man). Aquaculture Research. 2004;**35**:501-507

[20] Ortiz-Cornejo NL, Tovar-Ramírez D, Abasolo-Pacheco F, Mazón-Suástegui JM. Homeopatía, una alternativa para la acuicultura. Revista Médica de Homeopatía. 2017;**10**(1):28-34

Section 2

Sustainable Aquaculture

Chapter 2

Integrated Rice and Aquaculture Farming

Pamuru Ramachandra Reddy and Battina Kishori

Additional information is available at the end of the chapter

http://dx.doi.org/10.5772/intechopen.78062

Abstract

The burning problems like scarcity of food for ever-growing human population in the present world are addressed by adapting various methods for production of protein, carbohydrate, oils and other food materials. One of the methods to produce high amount of food is integrated farming including rice-aquaculture farming, which produces protein and carbohydrate as major components besides others. Rice-aquaculture farming produces grain (carbohydrate) and animal protein without affecting the quality and quantity of rice yield on the same piece of land and renders additional financial gain besides main crop (rice) like conventional monoculture. The aquatic species grown in the integrated culture are mainly distinct types of fishes, selected crustaceans and other selected species. Profitable rice-aquaculture integrated farming is popular in Asian countries than in Western countries. However, the integrated rice-aquaculture farming has its own limitations. The type of methods, culture species, influencing factors, and pros and cons of rice-aquaculture integrated farming are discussed in the present chapter.

Keywords: integrated farming, rice, aquaculture, fishes, crustaceans

1. Introduction

Ever-growing human population occupies the available land area in the world by encroachment. This is happening due to drastic increase in population dynamics worldwide. According to US Census Bureau world population estimate, the world's population in 1901 was 1.6 billion; it was 3 billion by 1960, 5 billion by 1987, 6 billion by 1999 and 6.8 billion by 2009 and became approximately 7.9 billion by 2018. Based on the statistics, it is known that one billion population is getting added to the current population by every 9–12 years. Currently, the second highest populated country in the world is India. Based on the recent statistics, India's present population has reached 1.2 billion. Though China is occupying the first place in highest population in the world, it is expected that India may become the highest populated country by 2050 as per World Population Data Sheet of 2009.

The size of the population depends on its number and the ratio of available bearable resources. The population size also is contingent on the way resources are used and disseminated between the populations. To meet the demands of forever growing human population, major agricultural fertile lands are getting converted into residential areas. The total available land area in India occupies 2.4% of total land area of the world. Added fertile land area exists in India than any other country except the USA and Canada, which are holding more water area [1]. Because of the availability of more arable land and water, India became an agriculture-based country. From ancient days, major Indian populations live in villages and revolve mostly around agriculture and allied activities. It is well exemplified with the statistics of Indian population since 1947; it is tripled due to improvised and sub-standard conditions in India [2]. However, due to enormous increase in population along with increased poverty, modernization and more effectively globalization and medication, majority of the adjacent villages of small towns and all most all large and moderate villages located around cities were urbanized. Urbanization caused drastic conversion of arable agricultural land into modern housing.

The food scarcity is the predominant factor that influences the growth and economy of the overpopulated country. Supplementation of nutritious food to the forever growing human population in a nation like India or overall worldwide is at most target of governments or World United Nations. Though the agricultural land area is getting reduced due to overpopulation in the world, it is the focus to produce sufficient or more amount of nutritious food to meet the demands of population. The conventional methods of agriculture may not produce or increase the food productivity. Alternative methods are in focus to produce high amount of food by using available agricultural land.

Since ancient days the traditional cultivation methods were contributed for the food and living safety throughout the world. But at the present day, farming is completely linked to high yield by usage of many varieties of pesticides. This type of farming certainly pollutes the environment and drags the farmer into debits due to high investment on crops in the case of crop failures, which ultimately cause suicides of the farmers and their families in developing countries like India. One of the reasons for this is majority of the farmers are focusing only on one crop at time where there is a high degree of uncertainty. Besides this, the usage of excess amounts of pesticides and fertilizers causes imbalance in ecosystem and alters the natural environment, which ultimately influences the flora and fauna of that area, including soil microbiota [3], which may cause drastic change in the livelihood and may also be one of the responsible factors causing severe change in natural cycle and seasons such as reduced rainfall, increase in the temperatures and short or prolonged yearly seasons besides health hazards in the consumers. To overcome the above circumstances, integration of various agricultural enterprises, viz., cropping, animal husbandry, fishery, forestry, etc. has great potentiality in the agricultural economy. The usage of expertise of conventional methods in combination of modern methods is well explained in many studies [4–6].

The conventional methods of agriculture are completely successful and are developed by farmers for different environments which maintain rich biological diversity [4–6]. According to

Csavas [7] the livestock integrated system was started in between fourteenth and seventeenth centuries by growers. The motivation overdue in integrated rice-aquaculture farming (RAF) is to diminish unused matter from several subsystems on the farm. The unused products generated during farming of subsystems were secondhand as contributions to other subsystems to progress the yield and minimize the cost of productivities for the several subsystems [8].

In this series, many methods were coming into picture to produce quality food. Best examples are the terrace farming, zero-acreage farming [9], mixed cropping, mixed farming, integrated farming, etc., which are in practice. The integrated rice-fish co-culture is not a new method of producing food; for centuries, it has been in practice and is designated as globally important agricultural heritage system [10]. Due to the importance of RAF, US Agency for International Development (USAID Mekong ARCC, 2016) encourages farmers in the Asian countries to continue the integrated farming with a refined form [11]. Due to its importance in many Asian countries including India, the present chapter deals with various aspects of integrated RAF.

2. Integrated rice-aquaculture farming (RAF)

Integrated culture is not new and exists naturally. The natural ecosystem itself is a big example for integrated culturing where a number of flora and fauna living together in the same area of land. Coming to the natural integrated system in the rice fields, besides fish species coming from outside through water, crabs grow by making burrows within the field. The best well-known natural example for integrated rice and aquatic species (ASp) is rice and freshwater crab culture. The freshwater crab *Oziotelphusa* (*Oziotelphusa*) *senex senex* enormously grow naturally in rice fields in India. Since its natural occurrence in the rice fields, this crab is popularly named as 'Indian rice field crab'. This crab is a good source of protein and is available throughout the year with no cost. The name 'poor man's protein' is also popular for this freshwater rice field crab since most farmers (poor villagers) consume it [12, 13]. Though the farmers do not focus on naturally occurring integrated rice systems like crab within the rice fields for commercial purpose, they understand the importance of more than one culture within the same field. One more thing that the farmers focused on is commercially important species and benefit out of its yield along with the main crop.

The culturing of ASp in the rice fields is originated from the farmers, who were experienced with fish along with prawns, crabs and other invertebrates grown naturally (ASp moves from ponds to rice fields through irrigated water) in the rice fields without adding any ASp seed in the irrigated field. The observation takes the farmers to culture ASp in the rice fields initially without any selection of the ASp. Later farmers are motivated to grow selected ASp with rice by following conventional methods based on the conditions of the local environment. In the ancient days, the farmers tasted the yield of integrated rice-ASp culture. In the beginning, it was called as rice + fish culture [14, 15]; later, it became rice-aquaculture. The uncontrolled entry of outside ASp into the integrated culture field and leaving the cultured ASp to outside the field are problems even today, but capture system of rice-ASp culture was introduced to overcome this problem [16]. In this system, cultured ASp purely depends on the naturally available feed in the field. But in general, farmers supplement feed for growing organism from outside. However, continuum in the transition from pure capture culture to a capture-based production system is gradual and is described by Halwart [17].

Integrated rice and aquatic species culture have lot of scopes and achievements of high productivity in terms of carbohydrate and protein with proper management methods and controlling systems. The selection of culture species plays a crucial role in productivity, besides environmental conditions and proper management at the time of culturing. However, the farmers are in practice to cultivate the integrated rice and aquaculture in many areas of Asian countries and some of the African countries.

2.1. Site selection

Certain criteria should be followed while selecting the site (land) for integrated rice-aquaculture. The selected area (place) should have an optimum rainfall of 80 cm in a year; the lands with uniform contour and high water retaining capacity of land are considered for site selection. The site selection for rice cum aquaculture farming is low lying area where water flows easily and is available at any time in needs. The fertile soil rich in organic manure and with high water holding capacity is used for rice fields. In general, soils with medium texture and loam with silty clay are most preferable for RAF [18].

2.1.1. Rice field selection

The topography and contours of the land will make the difference in preparation of rice field. There are three types of rice fields identified such as:

a. *Perimeter-type field*: the field in this type prepared with a moderate elevation and ground sloping on all sides into perimeter trenches which facilitates easy drainage. The middle of the land is the growing area for rice (**Figure 1**).

b. *Central pond-type field*: the pond is prepared in the middle of the rice field. This is the easiest way to produce trench for integrated aquaculture and rice production (**Figure 2**).

c. *Lateral trench-type field*: trenches were prepared on one or both sides of rice slopes (**Figure 3**).

Besides these three types, there are many types of trenches prepared for integrated rice-aquaculture. Different other types of trenches in the rice field are presented in **Figure 4**. The ideal ratio for making an integrated RAF is explained well, for example if a 1 hectare of land of integrated system is about 125 m × 125 m. In this, 0.67 hectares is the area to be used for rice, i.e. 102.5 m × 102.5 m, whereas 7.5 m × 440 m for four sides of the field which is equal to 0.21 hectares were used for aquaculture (fish, crustacean and other aquatic animal culture). About 0.12 hectares of area, i.e. 3.75 m × 485 m, will be used as embankment area. The other area of about 0.04 hectares is used for planting fruits.

2.2. Rice and aquaculture

RAF had been in practice since years due to the constant profits of integrated culture. However, the generally practised rice-aquaculture is categorized into three major types [19],

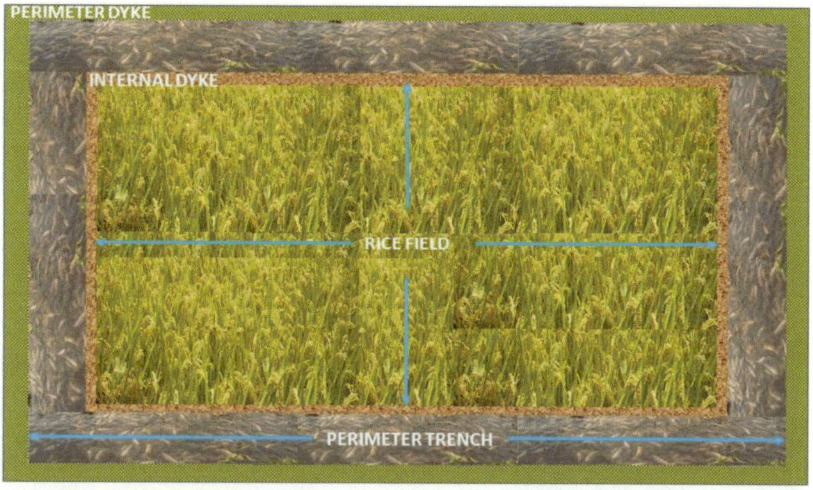

Figure 1. Perimeter-type rice field used for rice-aquaculture integrated farming.

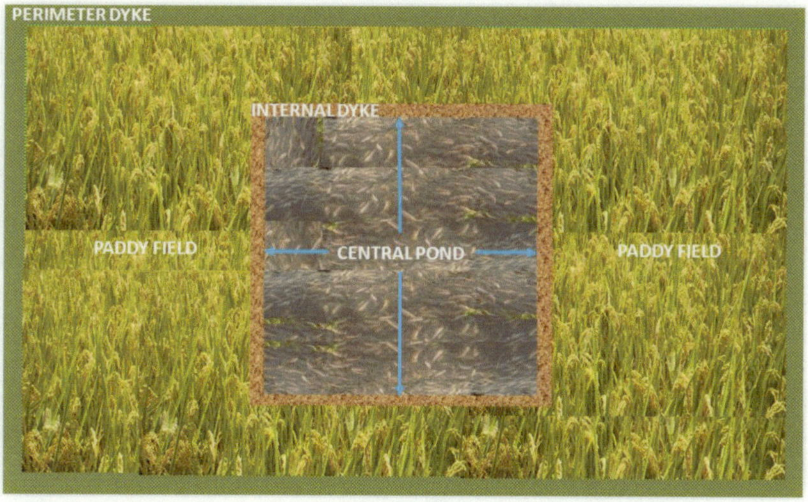

Figure 2. Rice field with central pond system for rice-aquaculture farming.

namely, (1) synchronous/combined/simultaneous farming, (2) sequential/rice-aquaculture rotation/alternate farming and (3) relay farming.

2.2.1. Synchronous/combined/simultaneous farming

In this type, the rice and aquatic culture species are grown simultaneously in the same field. Both the rice and cultured species are harvested at the same time. The synchronous rice-aquatic farming method has its own importance since it produces aquaculture protein with no additional cost; the cultured aquatic species secretes ammonia for rice growth besides destroying the weeds, tillering, enrichment of minerals (digging activity) and utilization of unused aquatic feed by the crop plant. Due to this the rice yield increases 5–15% than the conventional methods [19].

Figure 3. Lateral trench-type rice-aquaculture integrated farming.

This method of integrated culture is an old one and is familiar in the Asian countries. Usually in this type, 5 days after transplantation or after proper rooting of rice seedlings, the aquatic culture species seed is released into the trenches/pond (also called as stocking) [19]. The stocking time after transplantation of rice seedlings varies from aquatic cultural species to species. For example, stocking in the case of fry is after 10 days of transplantation, whereas for fingerlings after 3 weeks. The stocking procedure differs with the type of aquatic species growing (fingerlings or fish or prawn or crab or mollusk or any other aquatic species). In simultaneous farming, the rice selected is usually local variety harvested not less than 5 months by using organic manures like stocked cow/buffalo dong or green manure fertilizers without synthetic fertilizers. Moreover, the number of crops produced every year is two in this case. However, the combined rice-aquaculture farming has its own disadvantages such as high activity by predators of aquatic species (piscivore birds) due to less depth in trenches/ponds and restricted usage of herbicides and pesticides which causes harmful effects to growing aquatic species lowers rice yield; there may be 20–60% loss in aquatic species yield due to abrupt change in dissolved oxygen and temperature and limited space availability. Since having drawbacks in simultaneous rice-aquaculture farming, it was modified and established as alternative farming.

2.2.2. Sequential/rice-aquaculture rotation/alternate farming

Rice-aquaculture rotation is a simple method of farming where aquatic species were grown after harvesting of rice in the flooded fields without removing the rice stubbles. The water levels were shallow in the case of rice, whereas for aquatic species, water depth is raised. In this method more than one ASp were raised [19]. The depth for growing aquatic species in the field is raised well before the transplantation of rice seedlings with a moderate height of external dyke. However, the pesticides used for rice production degrades during interval between harvesting of rice and stocking of aquatic species. The rice stubbles in the water at the bottom of the field facilitate the growth of decomposing microbes which served as food

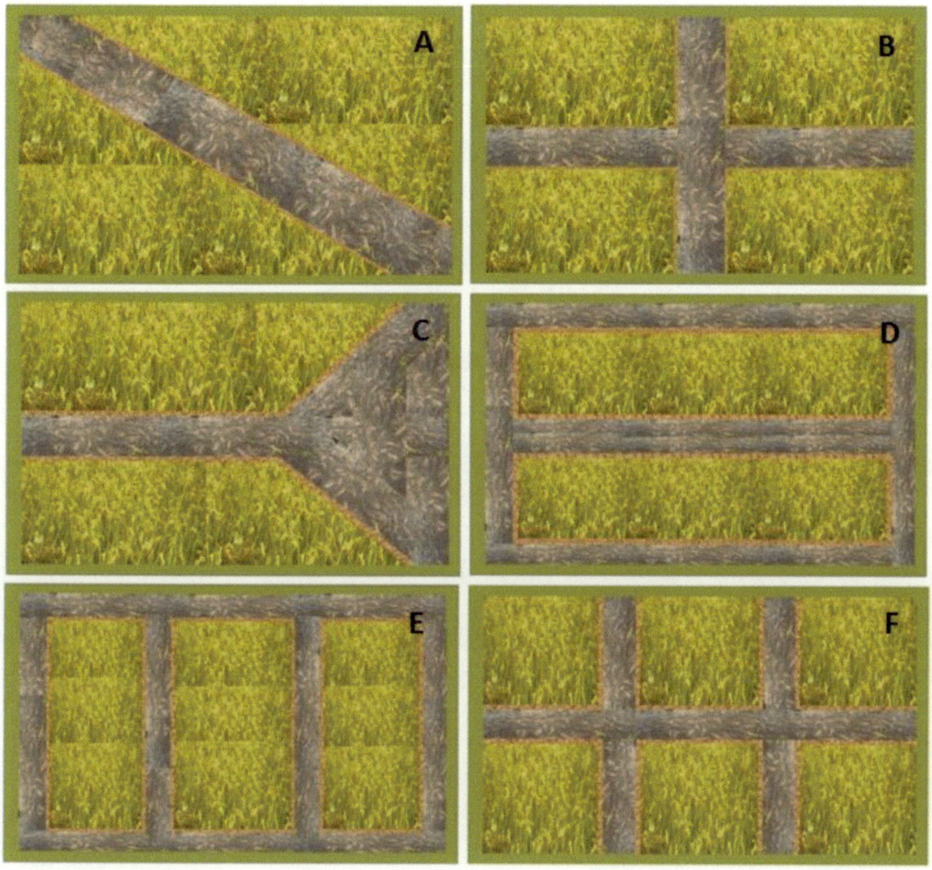

Figure 4. Different types of trenches in usage for rice-aquaculture integrated farming. (A) Diagonal trench, (B) crossed trench, (C) Y-shaped trench, (D) peripheral and one central longitudinal trench, (E) two equidistant transverses with peripheral trench and (F) latticed trenches.

for aquatic cultural species. The decomposition of stubbles enriches the water and soil with natural fertilizer utilized by the next cycle of rice where it helps to produce high yield.

The only limitation in alternate farming is about 20–60% loss of ASp due to piscivorous birds like herons and cormorants. Besides this, sequential rice-aquaculture farming has many advantages such as (i) no limitation for depth of water neither in rice cultivation nor in aquatic species culture, (ii) maintenance of adequate water levels provides sufficient dissolved oxygen and water temperature, (iii) rice stubble decomposition facilitates microbial food for aquatic species and fertilizer for the next cycle of rice for improved production, (iv) useful for mono- and polyculture of aquatic species and (v) reduced attack of insect pest on rice fields due to interruption in their life cycle in rotation farming.

2.2.3. *Relay farming*

The name 'relay' itself indicates it is a prolonged farming. The relay farming is a complicated system compared to synchronous and alternative rice-aquaculture farming and is a combination of these two methods [19]. In this farming rice and aquatic species were started like synchronous farming, but the aquatic species are not harvested while rice species are harvested. It means that relay farming requires longer period for aquaculture. During rice harvesting the growing aquatic species were transferred to special ditches/ponds which are connected to channels/pools of the rice field and then restocking them in the rice field after filling up of water for further growth. It means this method requires additional rearing facility for transferring fish at the time of rice harvesting. Relay farming provides high amount of aquatic protein with high yield of rice crop with short duration. The drawback in this method is more investment and labor. The carp and prawn species are the most commonly used to produce protein in relay farming.

2.3. Suitable aquatic species

Worldwide integrated farming of aqua with irrigation is becoming popular to produce more yield and makes the farmers economically strong. It is well practised in the rainfed areas mainly in countries producing rice. The 86% of rainfed areas in Thailand produce rice [20] as is the case with Lao PDR [21] and Cambodia [22, 23]. Various types of aquaculture in the rice fields are in practice. The major aquatic species that come under finfish and shellfish are the cultural species used for integration with rice fields. Among finfish and shellfish, the two types of cultures brackish and freshwater species have importance to grow in RAF. But the selection of this species purely depends on the variety of rice growing. It is very particular that only certain rice species are productive in brackish water. The rice varieties of holding salinity tolerance and floating-stem (long-stemmed) were preferred for brackish water RAF. The brackish water effect is tidal on rice species holding salt tolerance and long-stem, since these species have the capacity to withstand even in salt water. Besides rice variety the selection of fish or crustacean cultured also play a role in RAF. The freshwater prawns like *Macrobrachium* species and brackish water prawns like *Penaeus monodon* and other species grow well in the rice fields. Most species of fin fish can grow well in rice fields. However, the farmers have to be careful about the selection of aquatic culture species and the variety of rice for integrated culturing.

Studies were focused on the stability of rice field ecosystem with rice monoculture and rice-aquaculture [10, 24]. A survey by Xie et al. [10], with farmers cultivating rice with or without integrated aquafarming, revealed no difference in the yield of rice between the two farming for a period of 6 years. However, the temporal variation in the yield of rice is influenced by many factors such as year-round and year-to-year changes in climate, pest incidences, rice variety cultivation and the amount of pesticide used until harvesting [25, 26]. The usage of large amount of pesticides is responsible for temporal stability of rice yield in the case of rice monocultures, whereas rice yield stability is partly on the presence of aquatic species in rice-aquaculture farming [10], since fish-aquatic cultural species act as biocontrol agent [27]. In the modern agriculture, the stability of rice yield is maintained through the extensive usage of pesticides and fertilizers [28, 29]. In contrast, the stable yield of rice is highly maintained by exploiting synergies between species to minimize usage of chemicals and is suggesting to the modern agriculture that rice monocultures need to be improved with addition of species for facilitating positive interactions between species and components. The interactions between various components of RAF are presented in **Figure 5**.

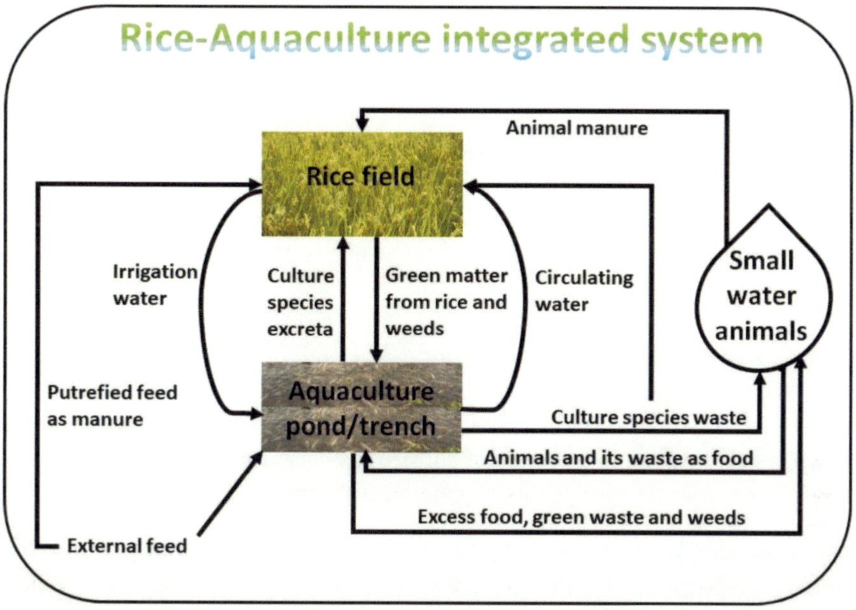

Figure 5. Interaction of different components in integrated rice-aquaculture systems.

The type of method for integrated culture mostly depends on local conditions of the culturing area. The climatic and other local conditions, cultured aquatic species and rice varieties, method of culturing, fertilizers used and supplementary feed are the principle factors that influence the quality and quantity of yield [30, 31]. The culture method selection depends on the source of aquatic species (ASp) and its interaction with the growing rice which was well described by Coche [31] seven decades ago.

The integrated rice-aquaculture is a mixed farming, where various rice varieties are grown along with selected aquatic cultural species depending on the local biological and physical factors. However, the selection of one aquatic species or more influences the yield of aquaculture protein. Because of this, the selection of aquatic species plays the most important role for the profit to the farmer and helps to improve the economy of the country. The simple economic benefit of shrimp/prawn, crab and fish with rice cultivation is well defined [32]. Besides profit rice-aquaculture generates the highest employability in rice-prawn culture, followed by rice-crab, rice-shrimp, rice-carp and rice-tilapia.

2.3.1. Suitable crustacean types

There are a number of crustacean species that grow in the rice fields as an integrated culture. The best example to explain this is the freshwater rice field crab that grows naturally in the rice fields of India. Besides crabs, there are other crustacean species like prawns, shrimps, lobsters and crayfishes that can also be grown in the integrated system along with rice with proper management [33], which is in practice in the coastal areas. Since the most profitable crustacean protein in today's world market is from shrimps or prawns, the production of these species is more profitable to the farmers especially the Indian and Asian conditions.

2.3.1.1. *Shrimp/prawn production*

Shrimps are grown in brackish water, whereas prawns grow in freshwater. The shrimp production is done with the fields of holding salty water located in west coast and deltaic areas of eastern India where the salt-resistant rice varieties are grown. Due to limited availability of salt water, only one crop (integrated rice-shrimp farming) is grown mostly in the months of July to September every year. The channels are arranged to control the water supply into the field or pond during harvesting of rice, and the shrimp is protected in the ponds/trenches. At this stage, rice field consists of substantial number of shrimp larvae having less water. The shrimp harvesting in this case is usually done after 3 months of rice harvest. To clearly say the method followed to culture rice and shrimp is through alternative system (semi-intensive monoculture) with few modifications and balanced diet supplied during the culture. The average production rate of shrimp per hectare in this method is 1440 kg. *Penaeus monodon* is the most common species grown in this type of culturing. *Penaeus merguiensis, Penaeus indicus* and *Metapenaeus ensis* are alternative species to grow in this type of system.

The most popular species of freshwater prawn grown in the integrated rice-prawn culture is *Macrobrachium rosenbergii* (*M. rosenbergii*). The monoculture of *M. rosenbergii* in the rice fields is in practice in North Kuttanad, where almost 248 hectares are cultivated integrated rice-prawn culture in a year [34]. Like rice-shrimp farming, the alternative system of culturing is adapted for rice-prawn culture (semi-intensive monoculture). The productivity of prawn monoculture with rice is around 700 kg/hectare. On the other hand, the polyculture of prawn (*M. rosenbergii*) along with fish (*Catla catla, Labeo rohita* and *Ctenopharyngodon idella*) in the rice fields is also in practice, where the productivity of prawn is 285 kg/hectare, which has almost two times less production of monoculture of prawn in the integrated system. The yield of *Penaeus monodon* with mixed cultures in the rice fields is about 2135 kg/hectare [35].

2.3.1.2. *Crab production*

More than two decades ago, the role of indigenous technology in the production of rice-crab culture is well described [36]. The crabs *Oziotelphusa senex senex* or *Parathelphusa hydrodromus* are commonly called as 'freshwater rice field crabs' since their occurrence is naturally in the rice fields. Besides this *Eriocheir sinensis* (Chinese mitten crab) and *Scylla serrata* (mud crab) are the major crab species used for rice-crab integrated farming. Mitten crab is a famous cultural species of China, whereas the mud crab is the most economic crab identified and grows throughout the world. Both these species grow in saline water and are cultured along with salt-resistant rice species in the case of integrated culture. *Scylla serrata* is popular in Asian continent. This crab culture is picking up in the Indian states such as Kerala, Tamil Nadu, Karnataka, Andhra Pradesh, West Bengal, Maharashtra, Odisha, Gujarat and Union territories. However, the crab culture is influenced by many factors, including environmental, biological, physical as well as socio-economic factors [37, 38].

Crabs depend on naturally available food in the rice fields. Anyhow, supplementation of external feed is needed during fast-growing periods. The management of rice-crab culture is much easier than other RAFs. But the only management aspect that needs to be addressed is regulation of pesticide usage. Since the usage of pesticides is unavoidable for rice yield, it can be managed by selecting less toxic pesticides with a method of spray on rice stems and specific time of spray at which crabs stay in burrows. Proper water management prevents the crab from escaping from rice fields. The average crab protein produced in rice-crab system is approximately 600 kg/hectare. The systematic management of crab culture changes the farmers socioeconomic status than the shrimp/prawn culture alone [39, 40] in addition, these crabs are more eco-friendly and grow easily in polyculture with other species [41, 42]. The only limitation in crab culture is limited availability of wild seed and is overcome by establishing crab hatchery industry to protect natural broodstock for continuous supply of seed [43]. The sequential or synchronous farming is the adaptable system for rice-crab culture, which depends on type of crab species growing.

2.3.1.3. *Crayfish production*

Similar to crabs, crayfishes are the strongest crustaceans that tolerate unfavorable conditions than other species. The most common species used in rice-crayfish integrated system are *Procambarus clarkii* and *Cherax quadricarinatus*. The rice-crayfish culture is common in Southern United States (*Procambarus clarkii*) and in Australia (*Cherax quadricarinatus*). The adult crayfish is used as broodstock to get seed in the ponds. The broodstock is released into field after 1 week of rice seedling implantation. On the other hand, the crayfish seed is also obtained from hatcheries. Besides this the seed also obtained from the rice-crayfish integrated system after harvesting, where the adult crayfish produce juveniles which are used as seed for immediate culture. Unlike other crustaceans rice-crayfish culture is also done with sequential farming. In this system before 2 weeks of harvesting the rice, the water is drained to facilitate crayfish to make burrows [44]. Watering the field after harvesting leads to growth of ratoon crop (occurrence of regrowth from rice stubbles) and facilitates the growth of insects, zooplankton, worms and molluscs, which promotes the direct growth of flushed out crayfish from the burrows [45]. The optimum weight for harvesting crayfish is about 15–60 g. The average yield of crayfish in this system is about 1960 kg/hectare [46].

Many other aquatic species can be integrated with rice culture. The typical integration of aquaculture with rice system followed in Asian countries is presented in **Figure 6**.

2.4. Rice-aquaculture farming with livestock

The profit of rice-aquaculture integrated culture created a scope for betterment of social-economic status of farmers. Using this as a concept, the farmers tried to integrate other livestock with RAF. There are many systems evolved in this way and mentioned in **Table 1**. However, the livestock rearing with RAF is in primitive stage and has its own limitations.

2.5. Management of rice-aquaculture farming

Individually the rice and aquaculture systems are separate entities but are mostly holding common requirements. Because of commonality in system management, the integrated RAF is prospective and profitable. However this integrated system has conflict in requirements and their management from one system to other. The modified interventions are required to overcome these problems at each and every aspect and step of system management. Besides selection of species (both rice and aquatic species) explained earlier, integrated system requires many other factors to manage and is explained hereunder.

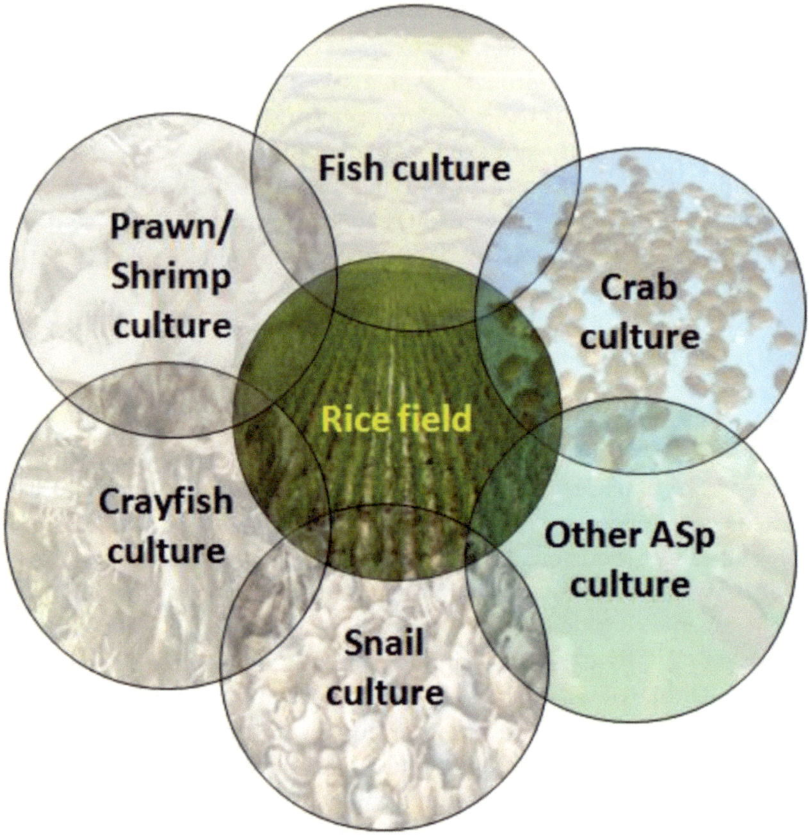

Figure 6. Integrated subsystems of traditional rice-aquaculture in Asia.

S. no.	Livestock integrated with rice-aquaculture system	Product produced from livestock	Reference
1.	Duck	Eggs and meat	Syamsiah et al. [47]
2.	Golden apple snails	Meat	Halwart [28]; FAO [48]
3.	Cows and buffalos	Milk and meat	Otubusin [49]
4.	Pigs	Meat	Prein [50]
5.	Sheep and goat	Milk and meat	Otubusin [49]
6	Poultry	Egg and chicken	Otubusin [49], Oladosu et al. [51], Momoh and Norman [52]

Table 1. The livestock integrated systems with rice and/or aquaculture in Asian countries.

2.5.1. Water management

Water plays an important role in any agricultural system. Continuous water supply is badly needed for RAF. The source of water (river, pond, well or bore) is the first parameter one should think of before establishing the integrated system. Uninterrupted water supply required for rice up to its maximum tolerated levels (15–20 cm) is recommended [53, 54] and is good for maximum yield of rice. For the aquatic species, the depth of water in the ponds/trenches should be 65–70 cm and is sufficient to manage cooler and hotter areas for the better growth of aquatic species. Most of the times, crustaceans grow at the bottom whereas fish at the middle and top layers of the water.

2.5.2. *Fertilizers and chemicals*

Organic and inorganic chemicals and fertilizers are essentially required for the growth of both rice and aquatic species. Besides nourishment of rice, external supplementation of reasonable amount of nutrients raised the levels of phyto- and zooplankton, which serve as food for aquatic cultural species. Aquatic species in the rice field enrich the soil fertility, which ultimately reduces the external supplementation of fertilizer and reduces the production cost [55]. This proves that RAF requires less usage of fertilizers than rice alone [56]. Alternatively yield of rice increases by the supplementation of nitrogen fertilizers at the time of land preparation instead of applying at the time of farming [57]. Addition of ammonia during farming causes damage to ASp as it acts as toxicant in the water, whereas no change has been seen with phosphorus fertilizer application in soil and water [56].

Organic fertilizers benefit for the growth of both rice and ASp. Moreover, the additional supplementation of animal manure helps for the better growth of aquatic species at the time of land preparation [58], and in this way, any toxic effects of these manure to the aquatic species are avoided [57]. Nevertheless, integrated rice-aquatic culture farming is good for poor and unfertilized soils where the aquatic culture species play the greatest role in making them fertile [24].

Besides fertilizers and nutrients, the control of pest and fungal diseases of rice is another task for production of higher yield of rice which usage is not at all good for the growth of ASp. In addition to reduced yield, the uncontrolled usage of pesticides and fertilizers declines the biodiversity. Prolonged misuse of pesticides and fertilizers over the years has also halted the development of inland fisheries and aquaculture [59–62].

In an attempt to reduce pesticide use, important changes have taken place in strategic approaches to plant protection. Integrated pest management (IPM) methods have brought ecological principles and social scientific perspectives into traditional crop management. These ecology-based pest control methods have resulted in markedly improved rice farming systems, which are not only higher yielding but also more sustainable [63–66]. In addition to this, increased adoption of rice-fish farming, with fish as a natural control agent of pest organisms, provides a promising alternative for further developing ecological sound management strategies of the rice field environment [61, 67–69].

2.5.3. *Nutrition and supplemental feeding of aquatic cultural species*

The rice ecosystem is rich with natural flora and fauna. Naturally the rice ecosystem itself is a rich nutrient supplement for aquatic cultural species. The phytoplanktons and bacteria are the first developers in the rice field and later weeds (macrophyton, benthos and detritus), and zooplankton will develop. All these are directly consumed by the aquatic species. Usually, farmers enrich the natural nutrients of aquatic species by supplementation of fertilizers. An alternative to have natural supplementation in RAF is by raising a fern azolla. Azolla is the best natural feed for fish, but not for crustaceans. Anyhow, this naturally occurring food is not sufficient for growing aquatic culture species. The consumption and utilization of natural field nutrients vary from aquatic culture species to species [24]. So, the external supplementation of nutrients is a must for attaining higher yield of aquatic protein.

The composition of external feed supplemented for crustaceans/other aquatic species is a mix of 40% of animal waste (fish, snail, clam and viscera tRAFh), 25% of plant waste (rice or wheat bran, vegetables, pumpkin, leguminous cakes and sweet potatoes) and 35% of terrestrial gRAFs and/or duck weeds. These materials are made as powder and used for making food pellets. But various types of external feed are in practice in many countries [70–73].

2.5.4. Temperature and dissolved oxygen

The fluctuations in the temperature and dissolved oxygen do not have much influence on the yield of rice, but it matters for the production of aquatic protein. In summer, the water temperature for ASp is regulated by rice plant shading during summer [26, 74]. Anyhow the low/moderate temperatures are best suited for good yield of ASp. To maintain adequate temperatures for the better growth of fish, prawn, carb, crayfish, snails and other ASp, high amount of water needs to be maintained continuously in the ponds/trenches/refuges. The only way to maintain temperature in the rice-ASp system is by controlled water system.

The dissolved oxygen is an important factor that affects the survival and growth of ASp, and in extreme conditions, it may lead to death. Unsystematic management of ponds/trenches leads to pollution [75], resulting in reduced dissolved oxygen. Basically the reduced dissolved oxygen changes the physiochemical properties of water, thereby pond/trench ultimately shows effect on growing fish. Most of the times, dissolved oxygen is reduced due to overloading of nutrients in the ponds/trenches and also causes increase in cyanobacterial bloom, which depletes dissolved oxygen and shows poisonous effects on ASp [76]. The increase in photosynthesis in rice due to increased aeration is recorded with the fish movement in the rice field [77, 78].

2.5.5. Predators and other factors

The birds are the predators for the ASp in rice-ASp system. Small birds may not cause damage to the system, but the larger ones reduce the ASp production to 20–60%. Aquavorous birds should be controlled by placing certain bird terrorizing signs or holdings in and around the rice-ASp system or making sounds to create panic and foreboding in them.

2.6. Pros and corns

a. Benefits of rice-aquaculture farming:

1. Usage of available land effectively.
2. Provides employment for village labors than rice alone.
3. One-time investment is good enough for making trenches/ponds/refuges.
4. Not needing of weeding in rice and feeding for Asp ultimately reduces the cost of labor.

5. Aquatic excreta and the waste generated during culturing provide organic fertilization which improves the rice yield up to 15%.
6. Reduced usage of external fertilizers, pesticides and fungicides ultimately provides pollution-free environment, thereby protecting the nature.
7. The additional production of quality aquatic protein for nourishing people suffering from protein malnutrition.
8. The ASp movement and foraging in the rice field increase the dissolved oxygen, thereby increasing photosynthesis and rice production.
9. Rice and aquatic protein produced are hassle-free and organic, which may reduce the occurrence of several diseases caused by polluted environment and food.
10. The external quench may provide place for growing organic onion, sweet potato, bean, ladies finger, tomato, variety of green leaves and other vegetables, which provide additional income to the farmer.
11. The wastelands or land not under usage for cultivation becomes fertile.
12. The wastage of nutrients supplemented for the rice are prevented, since the microalgae, the competitors for nutrients with rice in the field, are controlled by ASp which they swallow as food.
13. The reduction of rice yield up to 50% by the aquatic weeds is controlled by ASp.
14. The ASp fed on insects controls the pest effect on rice and increased yield.
15. The seed of ASp is produced during integrated system and is used for the next cycle of culture.

b. Limitations/disadvantages of rice-aquaculture farming:

1. Rice-aquaculture system is restricted only to the areas where surplus water is available.
2. Greater amount of water supply to RAF may increase the water scarcity for human usage and difficulties in getting potable water in the future.
3. Labor required more than rice culture alone.
4. Only specified rice species tolerant of deep water/saline water and low temperatures can adopt this system.
5. Some aquatic species uproot or eat the rice seedlings.
6. The area of rice field is reduced due to ponds/trenches/refuges constructed for ASp.
7. Additional amount of fertilizers and feeding costs more than rice culture alone.
8. The pesticides and herbicides are restricted to use and may reduce the production of rice.
9. Abnormal change in the temperature, low dissolved oxygen, inadequate place and occurrence of birds catching the ASp lead to 20–60% loss in ASp production.

10. Controlling water supply and floods is difficult. Floods may lead to loss of ASp.

11. Most of the cases, only local ASp can be adapted to this system, but not the demand-based ASp culture.

12. Usually the irrigated water that comes from other lands may contain pesticides/herbicides which may drop the production capacity of this system.

13. ASp culture is confined only to the rice growing seasons.

Due to several disadvantages of rice-ASp culture, it has gradually been discarded, and farmers are going for rotational culture. Whatever so, the advantages of this system are more prominent and make the farmers to stand socio-economically especially in developing countries. The governments need to initiate to publicize the importance of rice-ASp integrated farming and its impact on the poverty, society and ultimately country.

3. Conclusion

Rice is a main carbohydrate ingredient in the regular diet of approximately 3.5 billion people worldwide, and developing countries are producing more than 90% of the rice in the world [26]. It is suggested that combination of traditional/conventional agriculture and familiar methods such as biotechnology and other modern technologies [79, 80] available may provide high yield by minimizing the usage of pesticides and fertilizers, within the available land area, water and nutrients. Adaptation of new methodologies into traditional RAF may provide chance to meet the global demand for food besides protecting the environment by reducing the pollution.

Acknowledgements

Authors are very much grateful to Prof. P. Sreenivasula Reddy for his suggestions and encouragement for the successful completion of this chapter.

Conflict of interest

Authors declares no conflict of interest.

Author details

Pamuru Ramachandra Reddy[1]* and Battina Kishori[2]

*Address all correspondence to: reddyprbiotech@gmail.com

1 Department of Biochemistry, Yogi Vemana University, Kadapa, A.P., India

2 Department of Biotechnology, Sri Padmavathi Mahila Viswavidhyalayam, Tirupati, A.P., India

References

[1] Sarah LK, Skoet J, Raney T. The number, size, and distribution of farms, smallholder farms, and family farms worldwide. World Development. 2016;**87**:16-29

[2] Mathew S. India's demographic transition and economic growth: Implications for China and the balance of power in Asia. IJMRME. 2015;**1**(1):513-516

[3] Chagnon M, Kreutzweiser D, Edward AD, Mitchell Morrissey CA, Noome DA, Van der Suijs JP. Risk of large-scale use of systemic insecticides to ecosystem functioning and services. Environmental Science and Pollution Research. 2015;**22**:119-134

[4] Altieri MA. Linking ecologists and traditional farmers in the search for sustainable agriculture. The Ecological Society of America. 2004;**2**:35-42

[5] Zhu Y, Chen H, Fan J, Wang Y, Li Y, Chen J, Yang S, Hu L, Leung H, Mew TW, Teng PS, Wang Z, Mundt CC. Genetic diversity and disease control in rice. Nature. 2000;**406**:718-722

[6] Altieri MA, Nicholls CI. Biodiversity and Pest Management in Agroecosystems. 2nd ed. New York, USA: The Harworth Press Inc.; 2002. p. 275

[7] Csavas I. Regional review on livestock-fish production systems in Asia. In: Mukherjee TK, Moi PS, Panandam JM, Yang YS, editors. Proceedings of the FAO/IPT Workshop on Integrated Livestock-Fish Production Systems; 16-20 December 1991. Institute of Advance Studies: University of Malaya, Kuala Lumpur, Malaysia; 1992

[8] Edwards P, Kaewpaitoon K, McCoy EW, Chantachaeng C. Pilot scale crop/livestock/fish integrated farm. AIT Research Report 184. Bagkok, Thailand. 1986. p. 131

[9] Thomaier S, Specht K, Henckel D, Dierich A, Siebert R, Freisinger UB, Sawicka M. Farming in and on urban buildings: Present practice and specific novelties of zero acreage farming (ZFarming). Renewable Agriculture and Food Systems. 2014;**30**:43-54

[10] Xie J, Hu L, Tang JJ, Wu X, Li N, Yuan Y, Yang H, Zhang J, Luo S, Chen X. Ecological mechanisms underlying the sustainability of the agricultural heritage rice–fish coculture system. PNAS. 2011;**108**:E1381-E1387

[11] USAID Mekong Adaptation and Resilience to Climate Change (USAID Mekong ARCC): Development of Rice-Shrimp Farming in Kekong River Delta, Vietnam. 2016. http://mekongarcc.net/sites/default/files/rice-shrimp_report_amdi_final_eng_format_approved.pdf

[12] Ramachandra RP. Isolation and Characterization of Eyestalk Peptide Hormones Regulating Molting and Reproduction in the Crab *Oziotelphusa senex senex* [Thesis]. Tirupati: Sri Venkateswara University; 2005

[13] Kishori B. Involvement of Opioid Peptides in Regulating Carbohydrate Metabolism and Reproduction in the Fresh Water Crab, *Oziotelphusa senex senex* [Thesis]. Tirupati: Sri Venkateswara University; 2003

[14] Yunus M, Hardjamulia A, Syamsiah I, Suriapermana S. Evaluation of rice-fish production systems in Indonesia. In: De la Cruz CR, Lightfoot C, Costa-Pierce BA, Carangal VR,

Bimbao MP, editors. Rice-fish Research and Development in Asia. Proceedings of the ICLARM Conference. Vol. 24. 1992. pp. 131-138, 457

[15] Roger PA. Biology and Management of the Floodwater Ecosystem in Rice Fields. Los Banos, Laguna, Philippines: International Rice Research Institute; 1996. p. 250

[16] Coche AG. Fish culture in rice fields a world-wide synthesis. Hydrobiologia. 1967;**30**:1-44

[17] Halwart M. Ricefield fisheries and rice-based aquaculture—Underestimated and undervalued resources. In: FAO Inland Water Resources and Aquaculture Service. Review of the State of World Fishery Resources: Inland Fisheries. FAO Fisheries Circular No. 942, Rev. 1. Rome: FAO; 2003. pp. 36-47, 60

[18] Halwart M, Gupta MV. Culture of Fish in Rice Fields. FAO and The World Fish Center; 2004

[19] CIFRI Technology. Paddy Cum Fish Culture. West Bengal, India: Central Inland Fisheries Research Institute, Barrackpore; 1984

[20] Halwart M. Trends in rice-fish farming. FAO Aquaculture Newsletter. 1998;**18**:3-11

[21] Funge-Smith SJ. Small-scale rural aquaculture in the a Lao PDR FAO Aquaculture Newsletter. 1999; Nos. 22 and 23

[22] Guttman H. Rice field fisheries—A resource for Cambodia. Naga, The ICLARM Quarterly. 1999;**22**:11-15

[23] Balzer T, Balzer P, Pon S. Traditional use and availability of aquatic biodiversity in rice-based ecosystems—I. Kampong Thom Province, Kingdom of Cambodia. In: Halwart M, Bartley D, editors. FAO Inland Water Resources and Aquaculture Service. Rome: FAO; 2002. CD ROM ISBN 92-5-104820-7

[24] Tilman D, Hill J, Lehman C. Carbon-negative biofuels from low-input high diversity grassland biomass. Science. 2006;**314**:1598-1600

[25] Thomas MB. Ecological approaches and the development of "truly integrated" pest management. PNSA. USA.1999;**96**:5944-5951

[26] Frei M, Klaus BK. Integrated rice-fish culture: Coupled production saves resources. Natural Resources Forum. 2005;**29**:135-143

[27] Halwart M. Fish as Biological Control Agents in Rice: The Potential of common Carp *Cyprinus carpio* l. and Nile tilapia *Oreochromis niloticus* (L) [Thesis]. Germany: University of Hohenheim, Hohenheim; 1994

[28] Omer A, Pascual U, Russell NP. Biodiversity conservation and productivity in intensive agricultural systems. Journal of Agricultural Economics. 2007;**58**:308-329

[29] Tilman D, Cassman KG, Matson PA, Naylor R, Polasky S. Agricultural sustainability and intensive production practices. Nature. 2002;**418**:671-677

[30] Schuster WH. Fish culture in conjunction with rice cultivation. World Crops. 1955;**7**:11-14 67-70

[31] Yang Y, Zhang H-c, Xiao-jun HU, Qi-gen DAI, Zhang Y-j. Characteristics of growth and yield formation of rice in rice-fish farming system. Agricultural Sciences in China. 2006;**5**(2):103-110

[32] Salam MA, Ross LG, Beveridge MCM. A comparison of development opportunities for crab and shrimp aquaculture in South-western Bangladesh, using GIS modelling. Aquaculture. 2003;**220**:477-494

[33] Anonymous. Algicide for direct-seeded rice. IRRI Reporter. 1971;**2**(71):2-3

[34] Nair RR. Integrated fish farming in kuttanad: Problems and prospects. In: Proceedings of the Seminar on Integrated Fish Farming in Upper and Northern Kuttanad. 1994. pp. 1-8

[35] Ghosh A. Rice-fish farming development in India: Past, present and future. In: De la Cruz CR, Lightfoot C, Costa-Pierce BA, Carangal VR, Bimbao MP, editors. Rice-fish Research and Development in Asia. Proceedings of the ICLARM Conference. Vol. 24. 1992. pp. 27-43, 457

[36] Rajasekaran B, Whiteford MB. Rice-crab production in South India: The role of indigenous knowledge in designing food security policies. Food Policy. 1993;**18**:237-247

[37] Largo D, Ohno B, Ohno M, Critchley AT. Constructing an Artificial Seaweed Bed. Japan International Cooperation Agency (JICA): Seaweed Cultivation and Marine Ranching. Kanagawa International Fisheries Training Center; 1993. pp. 113-130

[38] Kingzet B, Salmon R, Canessa R. First nation's Shellfish Aquaculture Regional Business Strategy. BC Central and Northern Coast. Aboriginal Relations and Economic Measures, Land and Water British Columbia. 2002

[39] Cholik F, Hanafi A. A review of the status of the mud crab (Scylla spp.) fishery and culture in Indonesia. The mud crab. In: Angell CA, editor. A Report on the Seminar Convened in Surat Thani, Thailand; November 5-8, 1991; Bay of Bengal Programme, Madras, India. 1992. pp. 13-27

[40] Overton JL, Macintosh DJ. Mud Crab Culture: Prospects for the Small-scale Asian Farmer. Vol. 5. Malaysia: INFOFISH International. FAO Publication; 1997. pp. 26-32

[41] Chong LP. The culture and fattening of mud crabs. Info Fish International. 1993;**3**(May/June):46-49

[42] Chandrasekaran VS, Perumal P. The mud crab, *Scylla serrata* a species for culture and export. Seafood Export Journal. 1993;**25**:15-19

[43] Salam MA, Ross LG. Optimising site selection for development of shrimp (*Penaeus monodon*) and mud crab (*Scylla serrata*) culture in South-western Bangladesh. In: Proceedings of the GIS 2000 Conference; Toronto, Canada. 2000

[44] NAS. Making aquatic weeds useful: Some perspectives for developing countries. Report of an Ad Hoc Panel of the Advisory Committee on Technology Innovation, Board of Science and Technology for International Development, Commission on International Relations. Washington D.C.: National Academy of Sciences; 1976. p. 175

[45] Chien YH. Double Cropping Rice Oryza Sativa and Crawfish *Procambarus clarkii* (Girard) [M.S. thesis]. Baton Rouge LA, USA: Louisiana State University and Agricultural and Mechanical College; 1978. 85 p

[46] de la Bretonne LW Jr, Romaire RP. Crawfish Production: Harvesting, Marketing and Economics. Southern Regional Aquaculture Center Publication. SRAC No. 242; 1990. p. 4. http://www.aces.edu/dept/fisheries/aquaculture/pdf/242fs.pdf

[47] Syamsiah I, Suriapermana S, Fagi AM. Research on rice-fish culture: Past experiences and future research program. In: De la Cruz CR, Lightfoot C, Costa-Pierce BA, Carangal VR, Bimbao MP, editors. Rice-fish Research and Development in Asia. Proceedings of the ICLARM Conference. Vol. 24. 1992. pp. 287-293, 457

[48] FAO. Fish-Farming in Vietnamese Rice Fields Fights Golden Apple Snail Pest. News & Highlights. 1998. FAO Web Page: http://www.fao.org/NEWS/1998/980410-e.htm

[49] Otubusin SO. A proposed integrated livestock-rice-poultry cum fish culture in enclosure system. In: Proceedings of the Fifth Annual Conference of the Fisheries Society of Nigeria. Ilorin: Fish; pp. 319-326

[50] Prein M. Integration of aquaculture into crop-animal systems in Asia. Agricultural Systems. 2002;**71**:127-146

[51] Qladosu GA, Ayinla OA, Onuoha GC, Needom JG. Performance of *Clarias gariepinus* in polyculture with *Oreochromis niloticus* under the integrated broiler chicken fish farming. NIOMR Tech. Pap. No. 65. 1990. p. 22

[52] Momoh EJJ, Norman JE. Integration of fish with poultry and rice production. In: Olapada JO, Senessie SMB, editors. Sustainable Integrated Pond-based Aquaculture with Rice and Poultry Production Economic, Social and Environmental Assessment. 1st ed. Njala University Press. pp. 77-88

[53] Rosario WR. Rice-fish culture. In: Philippines (BFAR) Freshwater Aquaculture Extension Training Manual. Technology. Vol. III. Quezon City, Philippines: USAID-Bureau of Fisheries and Aquatic Resources; 1984. pp. 189-206, 248

[54] Koesoemadinata S, Costa-Pierce BA. Development of rice-fish farming in Indonesia: Past, present and future. In De la Cruz CR, Lightfoot C, Costa-Pierce BA, Carangal VR, Bimbao MP, editors. Rice-fish Research and Development in Asia. Proceedings of the ICLARM Conference. Vol. 24. 1992. pp. 45-62, 457

[55] Wu L. Methods of rice-fish culture and their ecological efficiency. In: MacKay KT, editor. Rice-fish Culture in China. Ottawa, Canada: International Development Research Centre (IDRC); 1995. pp. 91-96, 276

[56] Li X, Huaixun W, Yontai Z. Economic and ecological benefits of rice-fish culture. In: MacKay KT, editor. Rice-fish Culture in China. Ottawa, Canada: International Development Research Centre (IDRC); 1995. pp. 129-138, 276

[57] Singh VP, Early AC, Wickham TH. Rice agronomy in relation to fish culture. In: Pullin RSV, Shehadeh ZH, editors. In: Proceedings of the ICLARMSEARCA Conference on Integrated Agriculture-Aquaculture Farming Systems; Manila, Philippines. 1979. pp. 15-34, 258

[58] Sevilleja RC, Cagauan AG, Lopez EA, de la Cruz CR, van Dam AA. Component technology research in rice-fish systems in the Philippines. In De la Cruz CR, Lightfoot C, CostaPierce BA, Carangal VR, Bimbao MP, editors. Rice-fish Research and Development in Asia. Proceedings of the ICLARM Conference. vol. 24. 1992. pp. 373-384, 457

[59] Moulton TP. More rice and less fish: Some problems of the "green revolution". Australian Natural History. 1973;**17**:322-327

[60] Cagauan AG, Arce RG. Overview of pesticide use in rice-fish farming in Southeast Asia. In: De la Cruz CR, Lightfoot C, Costa-Pierce BA, Carangal VR, Bimbao MP, editors. Rice-fish Research and Development in Asia. Proceedings of the ICLARM Conference. Vol. 24. 1992. pp. 217-234, 457

[61] Halwarth M. Fish as biocontrol agents in rice. The potential of common carp *Cyprinus carpio* and Nile tilapia *Oreochromis niloticus*. Tropical Agroecology, Vol. 8. Weikersheim: Hoenheim University; 1995

[62] Abdullah AR, Bajet CM, Matin MA, Nhan DD, Sulaiman AH. Ecotoxicology of pesticides in the tropical paddy field ecosystem. Environmental Toxicology Chemistry. 1997;**16**:59-70

[63] Stone R. Researchers score victory over pesticide-and pests-in Asia. Science. 1992; **256**:1272-1273

[64] Settle WH, Ariawan H, Astuti ET, Cahyana W, Hakim AL, Hindayana D, Lestari AS, Sartanto P. Managing tropical rice pests through conservation of generalist natural enemies and alternative prey. Ecology. 1996;**77**:1975-1988

[65] Noda T, Loc NT, Van Du P. Rice pest management in the Mekong Delta. In: Xuan VT, Matsui S, editors. Development of Farming Systems in the Mekong Delta of Vietnam. Ho Chi Minh: JIRCAS, CTU & CLRRI, Ho Chi Minh Publishing House; 1998. pp. 271-287

[66] Huan NH, Mai V, Escalada MM, Heong KL. Changes in rice farmer's pest management in the Mekong Delta, Vietnam. Crop Protection. 1999;**18**:557-563

[67] Cagauan AG. Overview of the potential roles of pisciculture on pest and disease control and nutrient management in rice fields. In: Symoens JJ, Micha JC, editors. The Management of Integrated Freshwater Agro-Piscicultural Ecosystem in Tropical Areas. Brussels: Royal Academy of Overseas Science; 1995. pp. 203-244

[68] Dela Cruz CR. Role of fish in enhancing rice field ecology and integrated pest management. In: Proceedings of the ICLARM Conference. Manila: International Center for Living Aquatic Resources Management; 1994

[69] Gesa Horstkotte-Wesseler. Socioeconomics of rice-aquaculture and IPM in the Philippines: Synergies, potential and problems. ICLARM Tech. Rep. 1999.57,225. ISBN: 971-802-004-7

[70] Gupta MV, Abdul Mazid M, Islam Md S, Rahman M, Hussain MG. Integration of aquaculture into the farming systems of the flood-prone ecosystems of Bangladesh: An evaluation of adoption and impact. ICLARM Tech. Rep. 1999. 56. p. 32

[71] He QH, Sheng HQ, Xu GH. The use of supplementary feeds for semi-intensive fish culture practices in reservoirs in China. In: Mathias JA, Charles AT, Hu BT, editors. Integrated Fish Farming. Proc. Workshop on Integrated Fish Farming; 11-15 Oct. 1994 Wuxi, Jiangsu Prov. P.R. China: CRC Press; 1998. pp. 307-324, 420

[72] Darvin LC. Status of rice fields as hatcheries/nurseries for tilapia in the Philippines. In: De la Cruz CR, Lightfoot C, Costa-Pierce BA, Carangal VR, Bimbao MP, editors. Rice-fish Research and Development in Asia. Proceedings of the ICLARM Conference. Vol. 24. 1992. pp. 145-150, 457

[73] Chikafumbwa FJK. Use of terrestrial plants in aquaculture in Malawi. In: Pullin RSV, Lazard J, Legendre M, Amon Kothias JB, Pauly D, editors. The Third International Symposium on Tilapia in Aquaculture. Proceedings of the ICLARM Conference. Vol. 41. 1996. pp. 175-182, 575

[74] Wahab MA, Kunda M, Azim ME, Dewan S, Thilsted SH. Evaluation of freshwater prawn-small fish culture concurrently with rice in Bangladesh. Aquaculture Research. 2008;**39**: 1524-1532

[75] Asala G. Principles of Integrated Aquaculture. In: Olatunde AA. In Lake Kaiinji. In: Ayeni JSO, editor. Helmited Guinea Fowl (NMGP). 1994

[76] Pearl HW, Tucker CS. Ecology of blue-green algae in aquaculture ponds l. World Aquaculture Society. 1995;**26**:130-135

[77] Kunda M, Azim ME, Wahab MA, Dewan S, Roos N, Thilsted SH. Potential of mixed culture of freshwater prawn (*Macrobrachium rosenbergii*) and self recruiting small species mola (*Amblypharyngodon mola*) in rotational rice-fish/ prawn culture systems in Bangladesh. Aquaculture Research. 2008;**39**:506-517

[78] Mustow SE. The effects of shading on phytoplankton photosynthesis in rice-fish fields in Bangladesh. Agriculture. Ecosystems and Environment. 2002;**90**:89-96

[79] Gebbers R, Adamchuk VI. Precision agriculture and food security. Science. 2010;**327**: 828-831

[80] Tester M, Langridge P. Breeding technologies to increase crop production in a changing world. Science. 2010;**327**:818-822

Chapter 3

Are Aquaculture Practices Sustaining Our Goal to Restore Oysters (*Crassostrea virginica*)?

Gulnihal Ozbay and Stacy L. Smith

Additional information is available at the end of the chapter

http://dx.doi.org/10.5772/intechopen.78989

Abstract

Coastal areas are home to a wealth of economic and natural resources and are the most developed areas in the nation with fast increase in human population. Over 50% of the nation's population resides in 17% of the contiguous U.S. coastal areas. It is critical that consideration be given to the impact humans have on these coastal ecosystems and to the methods which are currently being utilized to enhance and restore these coastal habitats. In this chapter, we compare the status of the Eastern oyster, *Crassostrea virginica*, in two east coast estuaries: the Delaware Inland Bays, Delaware and Apalachicola Bay, Florida. Many ecological services, which are provided by oysters, such as their filtration, benthic and pelagic coupling, and habitat forming characteristics, have been extensively studied and discussed. Many regional economies in the United States of which the harvest of Eastern oysters was a major component, struggled with the collapsed fishery due to habitat limitation, water quality, sedimentation, parasitic diseases and other land use impacts. In response to these issues, oyster aquaculture has grown and is now a major part of the working waterfront where traditional wild oyster populations used to thrive. Research focusing on the ecological effects of oysters farm-raised with commercial aquaculture equipment is becoming more prolific as the industry moves away from a wild harvest fishery to a cultivated product. The oyster fishery may be recouped if the demand for oysters is supplied with oysters from aquaculture operations. Our primary goal in this chapter is to increase awareness about the potential benefits and some of the challenges facing the increased presence of aquaculture in these estuary systems.

Keywords: Eastern oyster, restoration, enhancement, population dynamics, oyster aquaculture, estuary health

1. Introduction

Coastal areas are home to a wealth of economic and natural resources and are the most developed areas in the nation with continuous increase in human population. Over 50% of the nation's population resides in 17% of the U.S. coastal areas. In light of these numbers, it is critical that consideration be given to the impact humans have on these coastal ecosystems and to the methods which are currently being utilized to enhance and restore these coastal habitats. There are various ways people use coastal areas for their needs. Shellfish aquaculture is one of the many activities people conduct.

In this chapter, we compare and contrast the health and status of the Eastern oyster, *Crassostrea virginica,* in two east coast estuaries: the Delaware Inland Bays, Delaware and Apalachicola Bay, Florida. Many ecological services which are provided by oysters, such as their filtration, benthic and pelagic coupling, and habitat forming characteristics, have been extensively studied and discussed. Oysters increase water clarity and quality by filtering sediments and algae, and removing nutrients such as nitrogen and phosphorous. The Eastern oyster was once a fixture of the local economies on the east coast however, combined effects from over harvesting, habitat destruction, and diseases such as Dermo and MSX have caused oyster populations to decline dramatically. Along with this decline in oyster populations, coastal lagoons in east coasts of the United States have been experiencing rapid development within the coastal watershed increasing eutrophication events. The once abundant oysters filtered algae and sediments, removed phosphorus and nitrogen, and played a vital role in the ecosystem that could help to counteract the increasing pressure on the watersheds. Many regional economies in the United States of which the harvest of Eastern oysters was a major component, struggled with the collapsed fishery.

In response to these issues, oyster aquaculture has grown and is now a major part of the working waterfront where traditional wild oyster populations used to thrive. In recent years, farm-raised oysters have become a more sustainable operation than commercial fishing. Oyster aquaculture has benefits beyond supporting human economies and diets. Oyster aquaculture can provide many of the same ecological services as oyster reefs, which are a valuable component of estuaries worldwide, serving as a unique habitat for many ecologically and economically important species. Research focusing on the ecological effects of oysters raised with commercial aquaculture equipment is becoming more prolific as the industry moves away from a wild harvest fishery to a cultivated product. However, there is a critical need to better understand the dynamics of local waters to enhance potential fisheries for both estuaries. The oyster fishery may be recouped if the demand for oysters is supplied with oysters from aquaculture operations. Our primary goal in this chapter is to increase awareness about the potential benefits and some of the challenges facing the increased presence of aquaculture in these estuary systems.

2. Temperate estuary "Delaware Inland Bays" characteristics and challenges

Delaware's 'Inland Bays' (DIB), similar to many of the coastal lagoons in the Mid-Atlantic region of the United States (U.S.), have been experiencing the impacts of chronic eutrophication and sediment erosion resulting from several decades of poor land use practices including housing development, agriculture and sustained nutrient input from within the surrounding watershed [1]. The cumulative impacts of these effluents from anthropogenic activities has degraded water quality and reduced the diversity and abundance of various species of aquatic life including fishes, invertebrates and submerged aquatic vegetation [2]. As a keystone species in estuarine bays, oysters provide important ecological services in these systems by filtering suspended particulates from the water column, increasing water clarity, and removing nutrients from eutrophic waters [3, 4]. Oyster reefs also serve as a valuable component of estuarine ecosystems, offering unique habitats for many ecologically, economically, and recreationally important species [2]. The bay degradation has led to the dramatic decline of the local oyster *Crassostrea virginica* populations since the late 1800s [4–7].

In response to the plummeting populations, 'oyster gardening' programs have taken root throughout the estuarine ecosystems of the Mid-Atlantic, including Delaware Inland Bays (*see* **Figure 1a**), in an effort to restore the native oysters for their ecological and commercial contribution to the health and viability of coastal estuaries. Many community-based estuary programs have involved volunteers to help rear larval oysters into healthy adults for reef restoration and it is no different in Delaware Inland Bays [5, 9, 10]. Volunteers living in the local communities surrounding the watershed in the Delaware Inland Bays place floating baskets of oysters at the ends of their docks to allow the filter-feeders a safe haven to grow from small, young spat into thriving adult oysters (*see* pictures in **Figure 1b**). Community members throughout southern Delaware are being given the unique opportunity to observe first hand many of the important ecological services provided by oysters and learn about the local watersheds.

With a shoreline of approximately 418 km, no part of Delaware more than 13 km from tidal waters, with Delaware Inland Bay consisting of three shallow coastal Bays: Rehoboth, Indian River, and Little Assawoman Bays. The combined surface water area of the three bays covers 83 square km with an average depth of 1.2 m. The Delaware Inland Bays (DIB) supports a small commercial hard clam and blue crab fishery along with weakfish, spot, bluefish, and Atlantic menhaden representing the majority of the commercial finfish catch in the Delaware Inland Bays and a variety of other commercially and environmentally important aquatic species [64].

Associated problems in those bays are similar to other Mid-Atlantic estuaries including eutrophication, high turbidity, sedimentation, periodic hypoxic/anoxic conditions, annual fish kills, low species diversity, and physical disturbances due to anthropogenic activities especially in the man-made canal systems. According to Delaware Inland Bays Estuary Program Report [11] and Chaillou et al. [1], approximately 80% of freshwater flow is from groundwater and the sandy, permeable soils of the watershed have led to widespread contamination of groundwater by nitrates in Delaware Inland Bays. Flushing rates may vary widely among the three bay areas, being as low as 1–7 days for Little Assawoman while those for Rehoboth and Indian River Bays may be as high as 80 and 100 days, respectively.

Delaware Center for the Inland Bays Report [12] stated agriculture as the largest use of land (32%) followed by developed/developing lands (22%), forested lands (17%) and wetlands and waters (16% and 12%) with significant loss of forest lands recorded in the watershed between

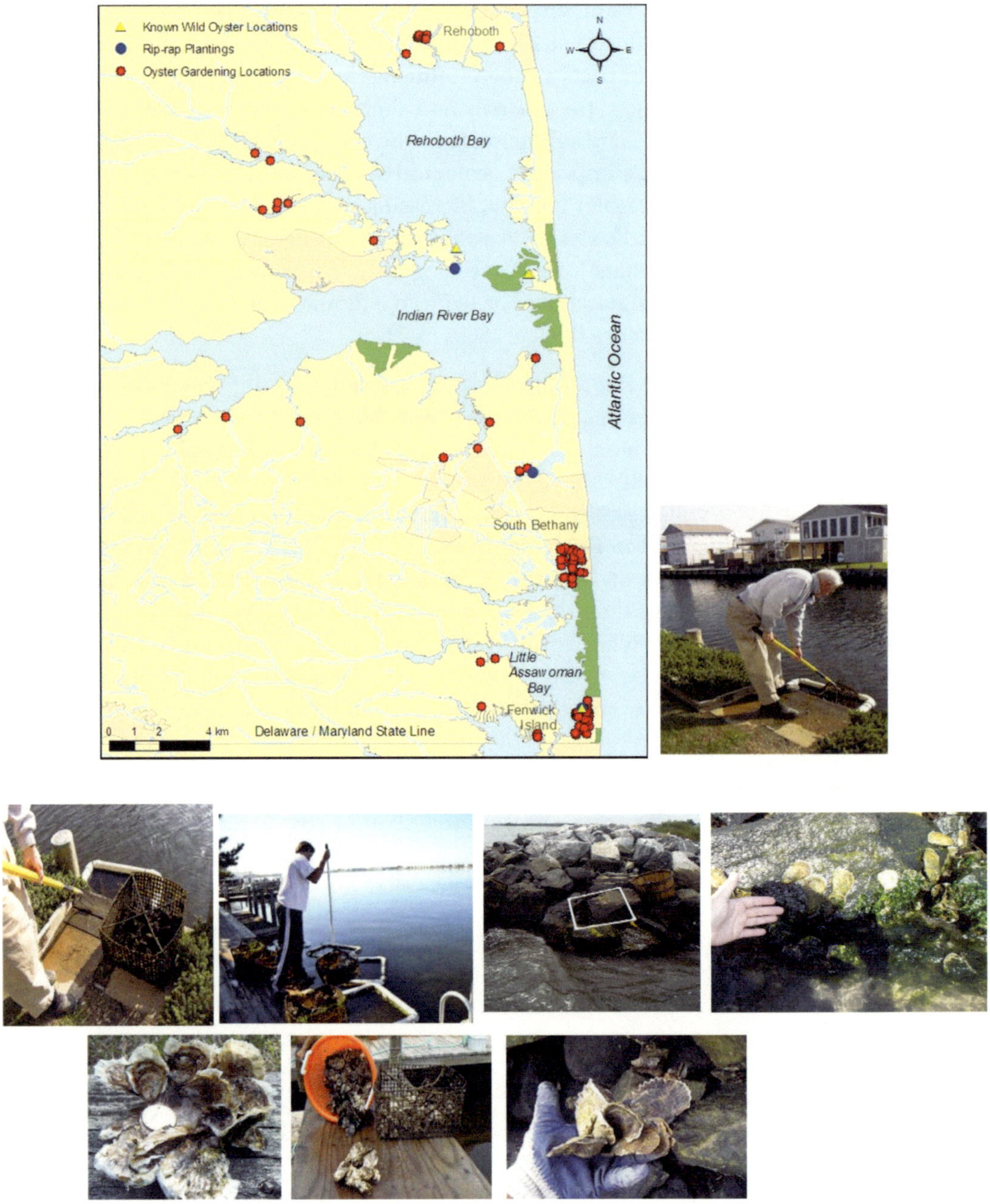

Figure 1. a. Delaware's Inland Bays showing oyster gardening locations, rip-rap planting locations, and known wild oyster locations [8]. *Map by Frank Marenghi.* b. Various oyster gears and oysters in rip-rap pictures indicating some natural recruitment is happening in Delaware Inland Bays. *Pictures by Frank Marenghi and Brian Reckenbeil.*

1992 and 2007 (*see* **Figure 2a**). **Figure 2b** shows the changes in the land use from 2007 to 2012, we can see improvement in the land use pattern for wetland lost. According to Delaware Inland Bays Estuary Program Report [11], the 200 hectares of dead-end canals within this system have been described as "unflushable." Martin et al. [15] reported that many of these canals are anoxic/hypoxic and subsequently lack higher trophic levels. **Table 1** describes the relative contribution of total nitrogen and phosphorus sources in the Inland Bays watershed and this outcome has not changed since the first time it was assessed in 1993 with nitrogen levels exceeding the targeted goal for all three bays in the Inland Bays. Agriculture is also listed as the leading contributor for the overall nitrogen and phosphorus sources in the Delaware Inland Bays [11].

Figure 3a shows the high nitrogen imputes, 6 times the healthy limit in Indian River due to fertilizer applications for agriculture and lawns in residential areas, animal waste and manure, and human wastewater [14, 16]. Eutrophication and degraded water quality impacts species present in this ecosystem [14]. **Figure 3a** displays the early nitrogen loadings in the Delaware Inland Bays from non-point source pollution [14]. This eventually causes regime shift from rich benthic flora and fauna to increase planktonic and microbial organisms [17]. **Figure 3b** shows the phosphorus loadings with no clear trends, according to Walch et al. [14], this may be credited to improved nutrient management on farms and the conversion of cropland to development with storm water controls.

Previous research suggested that [8, 18–24], Delaware's Inland Bays are in urgent need of the ecological services offered by oysters. Because these bays are very shallow (1 to 2.4 meter depth) and are poorly flushed by tidal movement, they are especially sensitive to environmental changes. Increases in pollutants, changes in salinity due to increase frequency of precipitation or drought events, climate change related fluctuation in water temperature, episodic hypoxic and anoxic conditions, as well as harmful algal blooms can all have detrimental effects on native oyster population. Proper site selection for oyster and reef restoration is essential and inclusive of other environmental limitation and issues. Over 50% of the available land in Delaware is used for agriculture with a long history of agricultural production of poultry, corn, soybeans, and other crops (*see* **Table 2**).

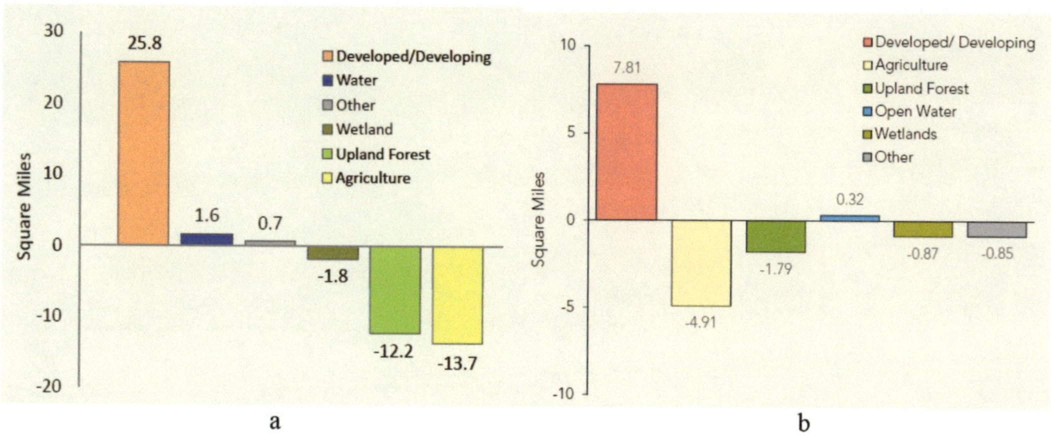

Figure 2. a. Changes in land use of the Inland Bays watershed from 1992 to 2007 [13]. b. Changes in land use of the Inland Bays watershed from 2007 to 2012 [14].

	Indian River Bay		Rehoboth Bay		Little Assawoman Bay	
Nutrient sources	Nitrogen	Phosphorus	Nitrogen	Phosphorus	Nitrogen	Phosphorus
Agriculture	44.6%	39.4%	33.0%	17.0%	54.7%	52.6%
Boating	< 0.1%	< 0.1%	< 0.1%	< 0.1%	< 0.1%	< 0.1%
Forest	11.0%	19.2%	7.4%	9.4%	6.7%	19.5%
Point sources	12.5%	15.0%	27.3%	56.9%	0.0%	0.0%
Rainfall	6.2%	8.6%	8.8%	6.9%	12.8%	11.5%
Septic tanks	16.0%	9.3%	11.2%	3.8%	14.6%	5.6%
Urban	9.8%	8.6%	11.7%	5.9%	11.2%	10.8%

Table 1. Relative contributions of nitrogen and phosphorus sources in the Inland Bays (Courtesy of [11]).

These adverse environmental impacts have detrimental effects on overall habitat quality and put tremendous pressure on local aquatic habitats. As Delaware's coastal landscape continues to develop in a low-density and sprawling manner, the health of valuable natural resources, many of which sustain local economies, is increasingly at risk. Managing the demands for protecting critical habitat areas and managing water resources are complex and continuous challenges in Delaware [26].

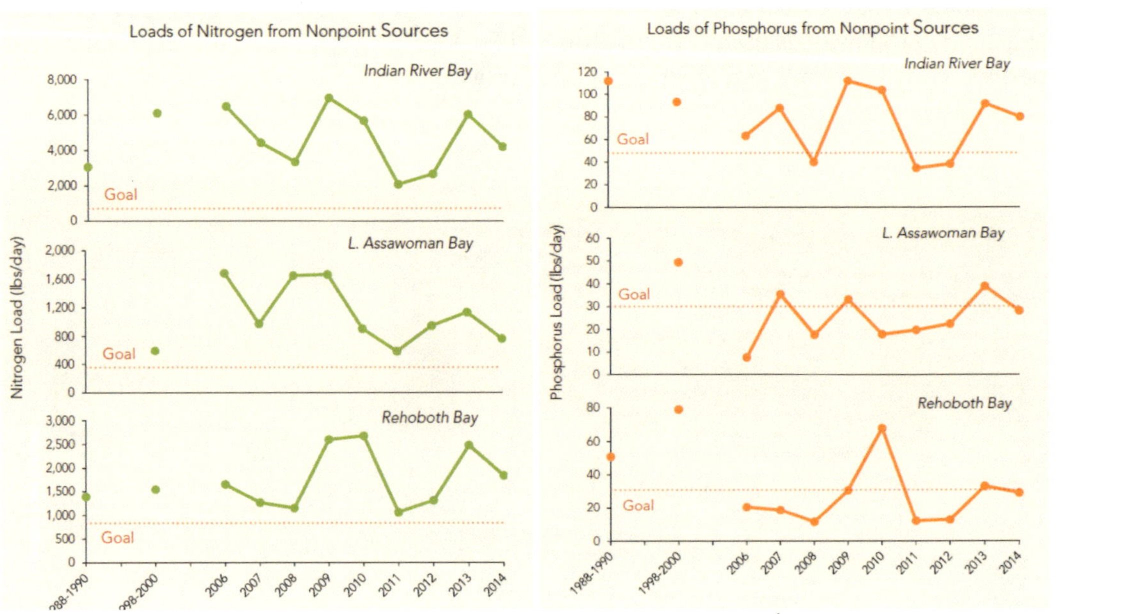

a b

Figure 3. a. Yearly nitrogen loadings in the Delaware Inland Bays from non-point source pollution [14]. b. Yearly phosphorus loadings in the Delaware Inland Bays from non-point source pollution [14].

According to U.S. Environmental Protection Agency [16], two areas of concerns have been identified as critical issues for DIB: eutrophication and habitat loss primarily due to urbanization, agricultural activities, and low flushing rates. Specifically, primary sources of nutrients include, a point and non-point sources in watershed, septic systems, animal wastes and fertilizers from agricultural lands. Excess nutrients, nitrogen and phosphorus deteriorated the bay aquatic life were managed using Total Maximum Daily Loads (TMDLs) for nitrogen and phosphorus for the Indian River and its Bay and Rehoboth Bay in 1998 and the Little Assawoman Bay in 2004. According to Delaware Center for the Inland Bays Report [27] "to meet the load reductions required by the TMDLs, water quality goals include the elimination of all point sources if nutrient loading to the water bodies, along with a 40% reduction in

Farms, Acreages, and Inventories

	Farms[1]	Acres
Corn for grain	843	183,000
Corn for silage	73	7,000
Wheat	340	80,000
Barley	137	35,000
Soybeans	817	170,000
Vegetables	233	42,200
Total Fresh Market Vegetables	187	11,400
Cabbage	6	490
Cantaloupes	42	300
Cucumbers	47	170
Potatoes	19	1,600
Pumpkins	36	800
Snap Beans	45	970
Strawberries	101	100
Sweet Corn	49	3,300
Tomatoes	82	300
Watermelons	86	2,700
Other Vegetables	7	670
Total Vegetables for Processing	89	30,800
Green Lima Bean	33	13,000
Sweet Corn	35	6,700
Green Peas	27	3,900
Other Vegetables	20	7,200
		Inventory
Broilers	778	51,092,495[1]
Cattle & Calves	399	18,000
Hogs & Pigs	77	5,000
Sheep & Lambs	56	903[1]
Equine[2]	2,000	13,000
Goats	232	3,530[1]
Bee Colonies	48	546

[1]2007, Census of Agriculture
[2]2004, DE Equine Industry Survey

Table 2. Delaware farms, their acreages, and types of farming practices [25].

nonpoint phosphorus loading in the Indian River Bay, Rehoboth Bay and Little Assawoman Bay, 65% reduction in the upper Indian River Watershed, a 40% reduction of nonpoint nitrogen loading in the Indian River Bay, Rehoboth Bay and Little Assawoman Bay, and an 85% reduction in the upper Indian River Watershed."

Figure 4 provides promising results in regards to reductions in point source pollution with five-fold decrease in total nitrogen and phosphorus concentrations from 1990 to 2009 was recorded in Rehoboth and Inland River Bays. However, relative concentrations of nitrogen and phosphorus from agriculture increased up to 57% from its previous levels of 45% and 39% for nitrogen and phosphorus, respectively [13].

Anthropogenic activities not only degrade water quality but they also contribute to the reduction in biodiversity and abundance of coastal bay species [2]. Eutrophication, high turbidity, sedimentation, periodic hypoxic/anoxic conditions, annual fish kills, low species diversity, and physical disturbances due to anthropogenic activities all contribute to reduction in biodiversity and abundance [2, 27, 28]. **Figure 5** provides a description of land use in the Delaware Inland Bays Watershed from 1992 to 2007. Significant increases are apparent in developed areas and areas marked for development. In opposition, declines were observed in areas that were upland forests or small agricultural areas [13].

To enhance habitat quality, for the past 15 years oyster gardening program initiated and public engagement has been the major part of the program effort to have the coastal citizens to be stewards of those bays and contribute to restoration efforts in the Delaware Inland Bays. Those oysters are further stored in the bays as adults with the hope that they will thrive in the natural setting. The resulting larger, healthier oysters are used for restoration work in the area such as artificial reef creation and rip-rap planting, and contribute spat to enhance wild populations. Although there is a general consensus among scholars that the current rates of resource depletions and environmental degradation cannot be sustained over a long period of time, these floating gardens are important in their abilities to offer essential habitat

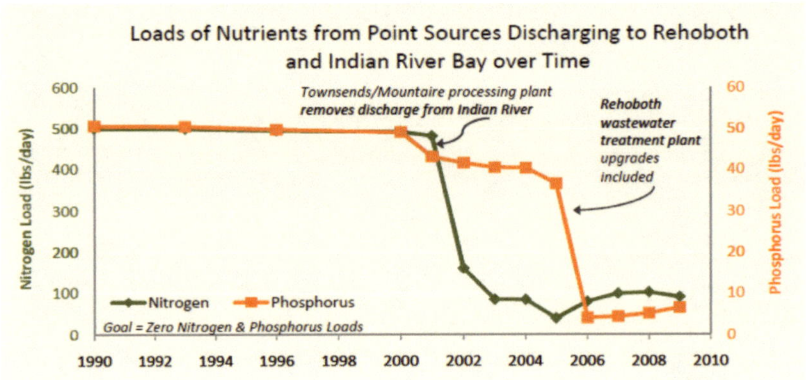

Figure 4. Nutrient loads reduction of point sources discharges in Rehoboth and Indian River Bay from 1990 to 2009 ([13]; www.inlandbays.org).

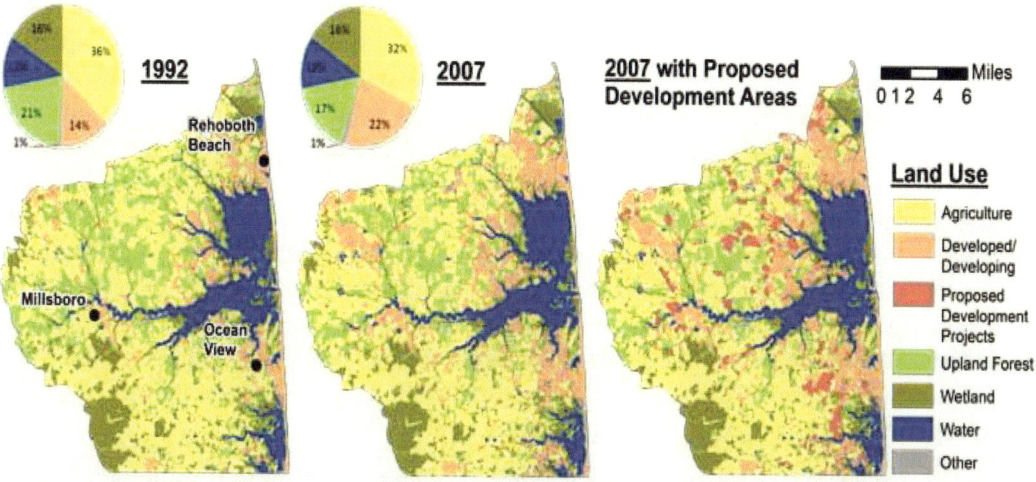

Figure 5. Changes in land use in the Inland Bays watershed [13].

revitalization. According to Kellogg et al. [29], oyster reefs reduce eutrophication by enhancing denitrification rates and assimilating nutrients into macrofauna.

Shellfish aquaculture has become a new hope for the coastal community in Delaware, with the approval of new regulations allowing commercial shellfish aquaculture practices. The past 10 years leading up to these regulations, Delaware Inland Bays have been home to a small community-based oyster mitigation program, which biennially distributes oyster spat on shell to volunteer citizen growers. The use of cost-effective culture techniques to culture oysters for restoration has developed into an integral part of the ecological restoration efforts.

Oyster gardening program was established to educate the public on the long-term stewardship and enhancement of the Inland Bays watershed as a collaborative effort with the leadership of the Center for the Inland Bays. The community oyster gardeners throughout the Inland Bays watershed support the program by caring for oysters held in floating cages 'Taylor floats' tied to their docks. Taylor floats are rectangular vinyl-coated 16 gauge, 25 mm wire mesh cages with a ring of PVC piping attached to the top to serve as the floatation gear (*see* **Figure 1b**). Each floating cage contains two square wire mesh baskets (46 x 46 x 23 cm) in which the oysters are placed [24].

Spat on shell were provided during the initial 3 years of oyster gardening program and later a remote setting process was implemented to supply the oyster gardening volunteers spats on shells. Remote-setting of oyster (*Crassostrea virginica*) larvae from the Northeast-Haskins resistant strain (NEH) line is performed biennially in Delaware to supply small-scale oyster enhancement efforts. Oyster larvae are raised in the flow through tank from the pediveliger stage through metamorphosis and settled on cleaned disarticulated oyster shells (cultch). Shell bags containing 5–10 mm spat have been distributed to oyster gardeners throughout the Inland Bays. In the floating cages, gardeners are able to keep the spat clean and protected, greatly minimizing the negative impacts of predators. Since 2009, simple alterations in choices of shell containment gear were made to try to increase settling efficiency rate and spat set. Shell containment gears included common diamond plastic mesh bags, wire baskets, and plastic aquaculture trays. Setting efficiency was estimated and for small-scale growers, the stacked aquaculture trays had the highest set efficiency and proved advantageous for several reasons, including: reduced handling time, uniform shell distribution within tanks, and easy-to-clean detritus between shell layers [84]. With improved growth and survivorship due to increased water flow, greater access to particulate foods, and much reduced risk of burial by sediments [9, 10], the resulting larger, healthier oysters have the potential to contribute spat for the enhancement of wild populations and are 'planted' in areas of the bays for local restoration work. Determining the remote set process success is often neglected, yet gathering this critical information will inform managers of the approximate number of spat distributed in small-scale programs and commercial scale aquaculture operation alike.

The creation of artificial reefs in designated areas is often used for oyster restoration, but the Delaware program started using riprap planting. Riprap is an irregular, large loose stones used to hinder the eroding effects of wave action. When oysters are planted in riprap, they are nestled in stable crevices between the rocks, mimicking the relatively secure, three-dimensional structure of naturally occurring oyster reefs that are integral to the oysters' survival [24]. Considering how limited oyster population in the Delaware Inland Bays, any effort to restore this keystone species in rip-rap crevices far closer than not making any effort and aquaculture is a step closer to a solution.

Although oyster aquaculture may be impacted by excess nutrients, it can also be a solution to mitigate this problem. According to Rose et al. [30], nitrogen removal by farmed shellfish was a more favorable solution per acre than BMPs for agricultural and storm-water run-off. Although new regulation allows oyster aquaculture in strategically identified areas in Delaware Inland Bays, Delaware is currently the only state on the Northeast Atlantic seaboard without commercial shellfish aquaculture. Legislation is developing policy and protocols for implementation, as the push for legalized aquaculture grows. Neighboring states have shown the economic and cultural benefits of functioning industry. Three Inland Bays in Southern Delaware, due to protection from open waters and ease of access for workers, offer promising future locations for bottom leases. Oysters are functionally extinct within the Bays and with the rapid development of the local watershed, the ecological services oysters contribute are more important than ever. Oyster aquaculture can help restore depleted wild populations of oysters while filtering the water, providing structural habitat, and creating a new sources of jobs. There is a unique opportunity to study directly how aquaculture facilitates restoration considering the impacts and benefits community driven oyster gardening program has provided since 2003. Delaware Department of Natural Resources and Environmental Control (DNREC), Delaware Division of Fish and Wildlife provides proposed shellfish regulations, proposed shellfish aquaculture development areas, legal notices and updates on regulations related information in their website at http://www.dnrec.delaware.gov/fw/fisheries/pages/shellfishaquaculture.aspx.

Figure 6 shows the shellfish growing areas in Delaware Inland Bays while **Figure 7** shows proposed shellfish aquaculture development areas in the Inland Bays.

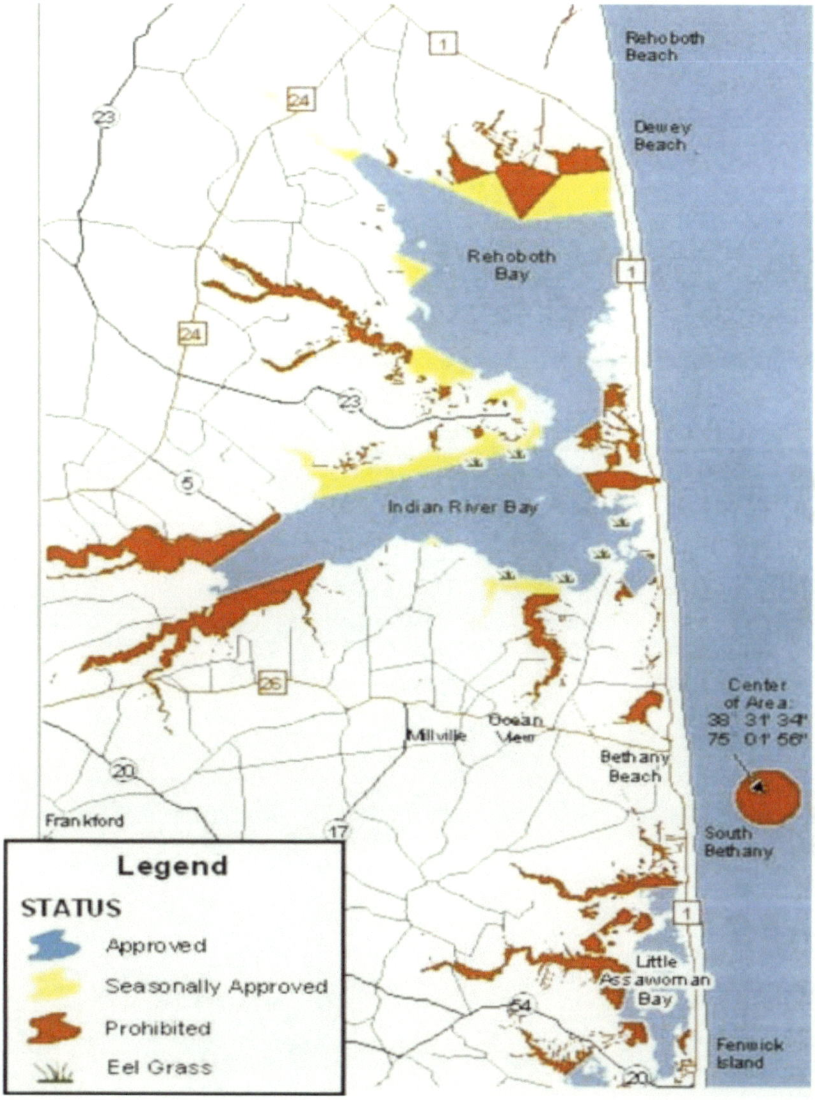

Figure 6. Delaware Inland Bays. Shellfish growing waters ([31]; http://www.dnrec.delaware.gov/swc/wa/Pages/Shellfish-Growing-Waters.aspx).

3. Subtropical estuary "Apalachicola Bay" characteristics and challenges

Apalachicola Bay is a subtropical, barrier island estuary located along the northeast Gulf of Mexico in northwest Florida. The bay, a National Estuarine Research Reserve (ANERR), is a river-dominated system [33, 34] with a highly variable salinity regime. Its main source of freshwater, the Apalachicola River, the largest river in Florida with the highest riverine discharge rate [35, 36] is formed at the confluence of the Chattahoochee and Flint rivers, both with

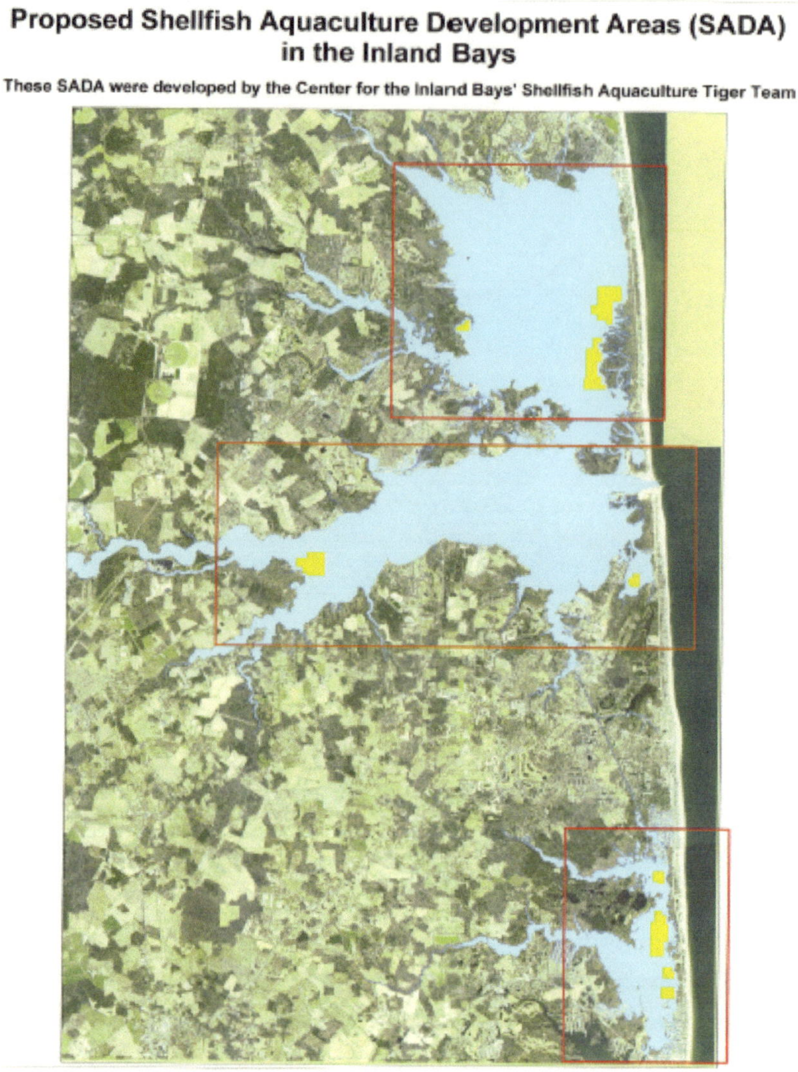

Figure 7. Proposed shellfish aquaculture development areas in the Inland Bays. ([32]; http://www.dnrec.delaware.gov/fw/fisheries/pages/shellfishaquaculture.aspx).

headwaters in Georgia. The Apalachicola-Chattahoochee-Flint (ACF) tri-river system drains 19,600 mi^2 of uplands and floodplains in Alabama, Georgia and Florida. The bay's hydrology consists of winter/spring flooding and summer/fall drought. The spring floods are essential to the health of the bay, which relies on Apalachicola River for freshwater and for the abundant nutrients – nitrogen, phosphorus and organic carbon – the delivers to the productive bay.

Species diversity is high in Apalachicola Bay, which has one of the most diverse ecosystems in the southeastern United States. Seafood production is a major industry in Franklin County, where the bay is located and where shellfish harvesting, especially of the Eastern oyster, *Crassostrea virginica,* contributed significantly to the local economy. The bay's oyster bars produced 90% of the oysters harvested in the state of Florida, and 10% of the nation's oysters and are known for their high quality and excellent taste. The industry generated \$10–\$14 million in revenue annually, and in Franklin County, oysters made up nearly one-third the value of commercial marine landings. **Figure 8** shows the interpreted surficial geology with the locations of oysters in Apalachicola Bay, Florida.

The river/bay ecosystems are in highly pristine areas and have not been adversely impacted by coastal development. The major stressor on the system, both the tidal and non-tidal reaches of the river, has been low river stage due mainly to three factors: dam installation, channel widening, and drought and natural fluctuations [36]. Over the recent past freshwater inflow to Apalachicola Bay has been critically diminished by the cumulative impacts of the aforementioned stressors, weather-related events and a decades-long water rights battle between Florida, Alabama, and Georgia (i.e. 'Tri-State River War'). Normal late-autumn drought conditions for the ACF watershed were exacerbated by two La Niña climate events in 2002 and 2007, during which time the southeast United States experienced warmer and drier than normal conditions. Drought-stricken Georgia increased its usage of the Apalachicola River's headwaters to support water demands from Atlanta's growing population and for crop irrigation. This resulted in a 17% reduction in water flow to the Apalachicola Bay. Other years experiencing unusually low river flow into the bay include 2000, 2008, 2011 and 2012 (the lowest on record); data show that the six lowest river flow years occurred between the years 2000–2012 [38]. The salinity in Apalachicola

Figure 8. Apalachicola Bay. Surficial geology shows the interpreted surficial geology with the locations of oyster bars superimposed on the sun-illuminated bathymetry ([37], U.S. Geological Survey Open File Report 2006-1381; https://cmgds.marine.usgs.gov/publications/of2006-1381/html/maps.htm).

Bay was exceptionally high in 2012 [39]. In January 2018, the Supreme Court of the United States heard arguments concerning the Alabama-Georgia-Florida water war. A ruling has not been issued at this time. Florida is seeking a water-sharing pact such that Georgia's usage of the ACF headwaters does not create adverse downstream effects for Apalachicola Bay fisheries.

In addition to decreased freshwater inflow, climate change models predict a north Florida sea level rise of up to 15 inches by the end of the century. Scientists speculate that this vertical rise may push the shoreline 70–250 feet inland in low-lying coastal areas (*see* **Figure 8**). According to a one report, this would submerge 61% of salt marshes and three quarters of the tidal fresh water marshes [40].

In August 2013, NOAA declared the Apalachicola Bay oyster fishery a disaster, caused by a long and excessive drought during the 2012–2013 season. Due to those events, Florida west coast oyster landings dropped 60% and revenue declined 44% [85]. The Deepwater Horizon Oil Spill did not impact Apalachicola Bay oysters significantly. Oysters tested by the University of Florida [39] were below instrumental detection for oil spill contaminants, polycyclic aromatic hydrocarbons (PAHs). According to research in the same report, a high percentage of bay oyster shells are parasitized by boring clams, sponges, polychaete worms or other organisms. In addition to a decrease in shellfish growth and productivity, shell deformity also detracts from shell integrity and may therefore affect the economic value of product. Dermo disease is present in Apalachicola Bay oysters, but apparently, its severity is less than in other bays along the East Coast, such as the Chesapeake Bay [41]. The UFL researchers report that more than 90% of tested oysters are positive for the parasite.

In an attempt to save a struggling industry, Florida's leaders have approved oyster and clam aquaculture leases in Wakulla County and in Franklin County (*see picture in* **Figure 9**). In April 2018, Florida's governor and his cabinet are looking to approve expanding current

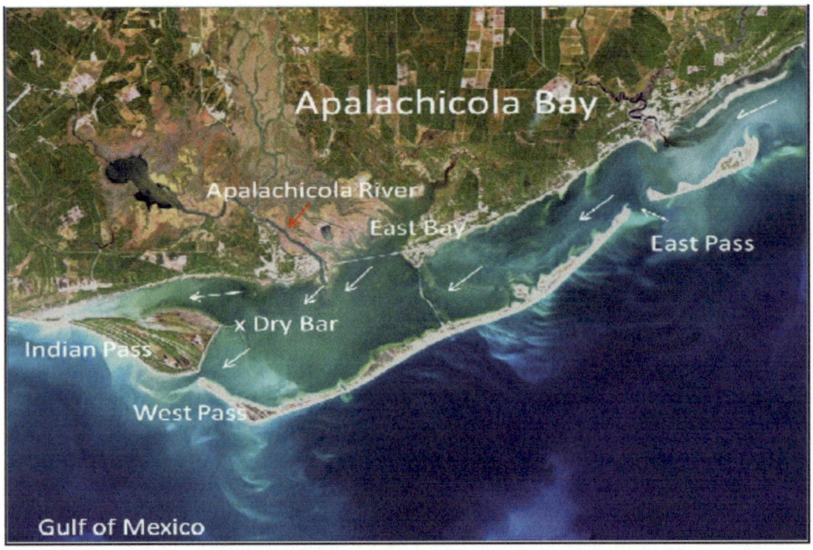

Figure 9. Apalachicola Bay relies on freshwater input from the Apalachicola River to maintain ecosystem health and to support a productive shellfish fishery. *Map from State of Florida, updated by Stacy Smith.*

Figure 10. A state worker showing FAMU students oysters in tongs in Apalachicola Bay. *Picture by Stacy Smith.*

aquaculture lease in Alligator Harbor, (21 new leases) Franklin, County, and Ochlockonee Bay (72 new leases), between Wakulla and Franklin counties. Each lease is 1.5 acres (The Apalachicola and Carrabelle Times 2018). Picture shows oysters in tongs in Apalachicola Bay and sediment (*see* **Figure 10**).

4. Where shared challenges meet shared solutions?

As stated by Rossi-Snook et al. [24] "an integral aspect of oyster gardening programs that cannot go unmentioned is the development of a sense of environmental stewardship among community members. In these programs, professional scientists and volunteers are working together to conserve both an ecosystem and a culture; by reintegrating oysters back into the bays, natural recruitment and proliferation is possible, eventually allowing for the safe and ecologically-sound harvest of oysters and other ecologically important macrofauna to redevelop within the community."

Ecosystem engineers, as described in many environmental books and articles, are organisms that can dramatically change the environment and essentially create ecosystems. Jones et al. [42] discussed differences between allogenic and autogenic ecosystem engineers. He stated oysters fall in to both categories: allogenic because they "change the environment by transforming living or non-living materials from one physical state to another, via mechanical or other means," and autogenic because they "change the environment via their own physical structures (i.e. living and dead tissue) as they grow and become larger, their tissues create habitat for other organisms to live in."

Although, *Crassostrea virginica* can tolerate a wide range of salinity, temperature, turbidity, and oxygen levels, Kennedy [6] discussed how water depth and salinity affect oyster populations and their associated fauna. Oysters generally occur in areas with the annual temperature range between −2 to 36°C except for the oysters in Gulf of Mexico which can survive intertidal temperatures between 44 and 49.5°C for over 3 hours. Larger established populations are found at salinities ranging from 5 to 40 ppt. Nevertheless, adult oysters have the ability to survive even in fresh water for short time durations [6]. When oysters are located in areas of an estuary with less salinity, they have slower growth rates. This is primarily due to a lack of food availability. In addition, because "drills, starfish, and boring sponges cannot stand the reduced salinities that prevail" in areas farther up in estuaries, oysters are able to have a higher rate of survivorship in these zones ([43], cited by [44]).

The Eastern oyster (*Crassostrea virginica)* serves as an essential connection between pelagic and benthic food webs. Oysters consistently remove suspended organic and inorganic particles >3 μm in diameter with much effectiveness [45]. Since it only takes up around 70% of the filtered organic material, this leaves dense, mucus-bound biodeposits, also known as pseudofeces which are ejected. These biodeposits can serve as a valuable food source for benthic organisms. Oyster reefs also prompt phytoplankton productivity in natural, non-eutrophic systems by vigorously filtering suspended materials, lowering turbidity that may restrict light penetration and oyster growth [45]. Increased water clearness will in turn promote growth of benthic algae and diatoms that are a substantial food source for sessile and mobile benthic herbivores that in turn are eaten by many carnivorous fish [3]. However, the function of oysters mineralizing organic carbon and converting nitrogen and phosphorus into forms usable by primary producers may be more critical than serve as the primary consumer in the salt marsh [46]. Oysters can enhance the reduction of ammonium to nitrites and nitrates through their biodeposition by taking N from the water column and depositing it into sediments. Microbes can then reduce the nitrogen to N_2 gas, which sublimates into the atmosphere. This is especially relevant in anthropogenically enriched environments [3, 46].

As [58] discussed oyster reefs have been considered as an essential fish habitat (EFH) for the last few decades. Many fish rely on oyster reefs for feeding, reproduction, and protection from predation. Within the same brief period during mid-summer, peak recruitment for all oyster reef residents occurs. This associates managing harvest and restoration efforts. Disruption of oysters by the addition of shell or dredging the reef during the spring through early autumn breeding season could negatively affect reproduction of many fish by burying nests, breaking apart articulated shells or scaring off males guarding their eggs [47].

Many economically important species may utilize oyster reefs for valuable juvenile nursery habitat Posey et al. [48]. Nursery habitat function of reefs may be expanded by locating restored reefs in shallow (<2 m deep) waters where large fish predators are less abundant. Important refuge habitats in estuaries are shallow water reefs. These reefs can also provide alternative foraging habitats for fish and crabs that are may be displaced by anoxic or hypoxic conditions as in the Chesapeake Bay, the Delaware Coastal Bays, the Gulf of Mexico and elsewhere [49].

Other nearby habitats can be influenced by oysters as well, like those of a salt marsh. This influence is achieved by protecting the salt marsh from the influences of wave energy. Shoreline retreat was significantly lower in a Louisiana study at sites with a constructed intertidal reef only 0.7 m tall in low energy areas. Low and high energy sites both showed positive oyster growth and recruitment (4.9 spat per shell) and showed potential to help stabilize sediment, reducing erosion, as well as providing salt marsh habitat in addition to a habitat of its own [50]. Because oyster reefs in salt marshes trap sediments as they grow, they can eventually become colonized by *Spartina* spp. and other grasses. Subsurface or fossil oyster reefs have been discovered extending from an existing reef into the marsh [46].

Indeed, bivalves can affect a shift in the phytoplankton community. Many authors have suggested that bivalves exert a "top-down" control of phytoplankton dynamics [51–54]. Some have stated in certain instances there exists a synergistic feedback. Oysters considerably advance the timing of nutrient recycling rates, allowing indulgence consumption and fast growth of some algal species [55, 56]. Through continuous filtering activities of the oysters, the phytoplankton community is shifted from older lag phase cells to younger cells in a logarithmic growth phase. Reduced competition causes the phytoplankton community to shift to faster growing algal species that are able to take advantage of the increased light and nutrients [57].

Crassostrea virginica are bivalve suspension-feeder organisms that are native to the Delaware Inland Bays and Apalachicola Bay areas. Bivalve suspension feeders have been shown to serve an important biogeochemical role in coastal ecosystems because N and P from the water column are transferred to the sediments in their biodeposits [3]. This means that the filtration of the oysters can be shown to remove nitrogen from an aquatic ecosystem. Although the benefits of physical oyster structure may be significant, the benefits from oysters' ecological function are under-appreciated [58]. Although many studies have been done to focus on various effects of oyster ecology on lower trophic levels, resident species, and water quality, very few studies have yet conclusively demonstrated net benefits to higher trophic levels.

One of few studies showed where oysters enrich the surrounding benthos with their biodeposits and dissolved nutrients increases meio- and macrofauna species assemblage in their study [46]. Also observed by Bahr and Lanier [46], many oyster reef residents feed upon these lower trophic levels and find aid in the unique niches created by the oysters. Polychaete were collected from the sediments under the oyster gears, 1 meter away from the gears and 5 meters away from the gear in the Delaware Inland Bays. Polychaete survey results during the warmer months indicate the highest abundance of polychaetes were found at the Little Assawoman site and the lowest abundance of polychaetes in Rehoboth Bay. The results of the benthic community assessment indicate that there was no significant impact to Polychaete abundance or species richness from the oysters and aquaculture gear. Little Assawoman Bay had higher abundance and species richness than other two bays [59]. Benthic community assessments are often used to evaluate the health of an ecosystem. A healthy benthic community in the mid-Atlantic is characterized by high biodiversity of benthic flora and macrofauna [17]. Benthic communities are made up of a several different types of organisms including many invertebrate species [86]. Benthic organisms play important roles in ecosystems because they are a fundamental part of the food web. They act both as a food source for larger organisms and as decomposers, helping bacteria break down organic matter [86].

Improved water quality and continuity of a healthy food chain, benthic/pelagic coupling, and planktonic stability by oysters have provided valuable benefits to estuaries [60]. Posey et al. [48] examined whether resident and transient species are in fact attracted to the physical structures of the oysters for feeding or if they receive the majority of their foods elsewhere and actually utilize these habitats created by oysters for other needs.

In addition to their impacts as a filter feeder to clarify water, harvested oysters left unharvested would remove excess nutrients from the bay incorporated within the oysters shell and tissue [3]. Floating aquaculture gear would increase species diversity in the Delaware Inland Bays by providing refuge and foraging areas for transient species moving throughout the Delaware Inland Bays [21, 24, 61]. The years of research efforts conducted by the primary author and her research team found [62, 63], unlike some finfish farming, rearing shellfish in high densities in shallow water can have positive effects on the environment and may promote biodiversity. In the Delaware studies conducted around the submerged aquaculture equipment, 17 species showing significantly greater abundance and richness than in adjacent low-profile oyster shell reefs in 2006. Fourteen species around the equipment vs. the eutrophied, turbid, soft-bottom lagoon (including 3 species that require oyster shells for spawning substrate) in 2007. About 49 species of fish and invertebrates along with 8 species of macroalgae greatly contributing to the diversity of the native ecological community in 2008. In Virginia, 45 species of macrofauna were recorded inhabiting one commercial oyster farm that used floating equipment. In a study in Rhode Island, species richness was significantly greater in submerged aquaculture equipment than in a nearby seagrass bed or an unvegetated sand flat, especially for fishes and invertebrates in their early life stages, demonstrating the equipment may benefit some species more than others. These studies are critical to understanding the complex ecological interactions that occur and will allow farmers, managers, and regulators to fully appreciate the consequences of their actions. **Figure 11** shows the aquaculture gears used for the oyster gardening program and previous studies.

The potential effect of utilizing shellfish aquaculture for community-based restoration and environmental conservation is promising. **Figure 12** shows 2011 shellfish harvesting status of the Delaware Inland Bays [64]. Suitable locations for spat recruitment and oyster growth can be used to advance natural oyster settings. Number of oyster gardeners currently involved in the Delaware Inland Bays (DIB) oyster restoration efforts is about 200 community volunteers using their docks. Working with this number, and that fact that each oyster filters approximately 190 liters of water per day, the oysters currently involved in the program filter about 7,570,825 liters of water per day in the Delaware Inland Bays. Although this may seem to be an impressive amount, it is not when observing the actual volume of the Delaware Inland Bays. The Delaware Inland Bays have a surface area of 83 square kilometers, with an average depth of 1.2 meters [15]. This is a total volume of 101 billion liters. In order to filter the volume of water in the Delaware Inland Bays once daily, at least 534 million more oysters need to be cultivated and allowed to live without harvest. There are currently about 40,000 oysters. Once the critical amount of at least 534 million oysters is established, only then will there be at a point where there will be excess for actual harvesting. Only just beginning to touch the tip of the proverbial iceberg in Delaware with the restoration project, many more efforts are required.

Habitat restoration and major pollution reductions are needed to restore water quality and achieve a healthy estuary once again. Unfortunately, areas close to the shoreline and most tributaries have unhealthy oxygen levels with severe condition in some areas although most open water areas have good dissolved oxygen for healthy aquatic lives. One of the major causes for poor water quality condition for low dissolved oxygen is due by the excess nutrient leading major habitat loss and degradation issues for variety of finfish, shellfish and other aquatic species including invertebrates [64].

Figure 11. Aquaculture gears used during the oyster gardening program. *Pictures by Frank Marenghi and Patrick Erbland.*

According to Delaware Center for the Inland Bays Report [64], the Delaware Inland Bays are a premier east coast fishing destination, but important state fishes like the weakfish and blue crab population are declining. While the Inland Bays Oyster Gardening Program and student research projects confirm oysters can grow successfully in all three bays, wild oysters are very limited.

Fulford et al. [65] shares findings suggest that the ecological benefit of restoring bivalve populations are somewhat variable. A comparative model analysis of restoration plans in specific systems can be highly beneficial to maximizing the benefit-to-cost ratio of restoration efforts intended to reduce the negative effects of cultural eutrophication. It should also be noted that we should be cautious of generalizations about the effect of suspension-feeding benthos on phytoplankton without due consideration of estuarine size, circulation patterns, and morphology, as well as any other factors that may regulate community filtering

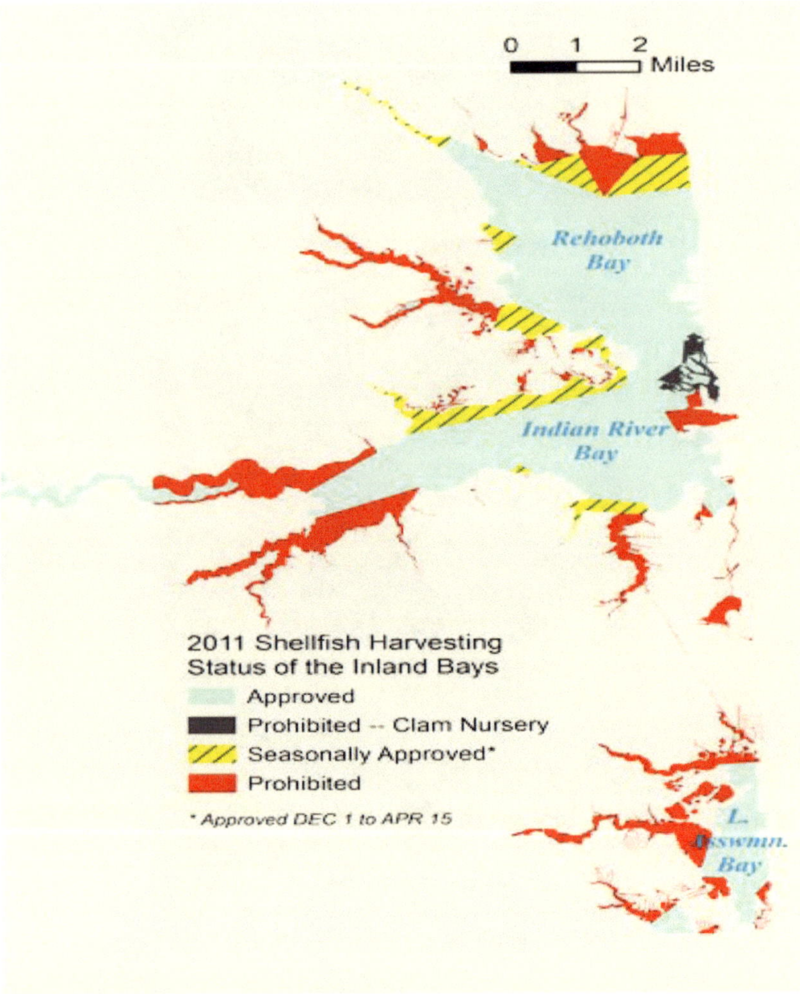

Figure 12. 2011 shellfish harvesting status of the Delaware Inland Bays (Delaware Center for the Inland Bays 2013).

rates [66]. Keeping these thoughts in mind for future research potential, let us conclude with a short synopsis of how well oyster gardening is working in the DIB, and what steps need to be taken nest to maximize enhancement and restorations of both temperate and subtropical estuaries.

For the past 15 years, oyster gardening has been part of the restoration of the Delaware Inland Bays. Volunteers living in the local communities surrounding the DIB place floating baskets of oysters at the ends of their docks, allowing them protection from predation in order to grow from small, young spat into thriving adult oysters. Oyster aquaculture has a potential to generate income for coastal communities [67]. The volunteers who participate in the monitoring of water quality and oyster aquaculture have learned many things while becoming trained in aquaculture. They have realized an increase awareness to protect our water and its biota. They have also gained an appreciation for the Eastern oyster species, *Crassostrea virginica*, and the efforts of its restoration. Oyster culture has a potential to lessen pressure on natural overexploited populations and to generate income for coastal communities [67]. There are various social and economic benefits of oyster gardening in their local habitats in relation to watershed improvements discussed by Ozbay and Cannon [68]. The implications of these gardeners' actions are exponential in their ability to offer essential habitat revitalization.

Environmental factors determine the productivity of the Apalachicola Bay oyster community: factors which encourage oyster growth include bottom substrate, nutrients from Apalachicola River and food availability. Those which are detrimental to oyster productivity include predation, disease, and sedimentation [69, 70]. Approximately, 10% of the bay's aquatic area is covered by oyster bars. Apalachicola Bay supplies the Florida seafood industry with 90% of its oysters. Local oyster harvesters and seafood suppliers rely on oysters for their livelihoods. The oyster industry brings $10–$14 million in revenue annually to Franklin County, FL; therefore, oyster productivity is linked to both ecosystem health and to financial solvency of the local economy.

Approximately 17% of the bay's total area is occupied by fresh, brackish and salt water tidal marshes and only 7% of its area is occupied by seagrass, with the majority of these seagrass beds confined to high salinity and low turbidity regions of the bay [71]. The riverine discharge and associated seasonal flood-related flux of inorganic and organic nutrients into the bay from the Apalachicola River and its associated marsh systems is essential to the present ecosystem dynamics of the estuary [33, 72]. The bay also supports highly productive shellfish and finfish fisheries all of which are either directly or indirectly dependent upon the hydrologic conditions of the bay. For example, when the salinity of this estuary increases, oyster mortality rates increase due to predation by Gulf of Mexico gastropod mollusks and other predators, which require higher salinities [71, 73, 74]. Input of freshwater in river-dominated systems reduces predation pressure from marine species during high flow periods.

A study by Chanton and Lewis [72] compared ecosystem biogeochemical dynamics in Apalachicola Bay during periods of low river flow (summer-autumn) versus high flow (winter–spring). They demonstrated that floodplain detritus does not drive estuarine production in Apalachicola Bay but rather the highest estuarine productivity in Apalachicola Bay coincides with low flow period during the summer [72, 75–77]. The bay's primary productivity during these low flow periods, however, is driven by autochthonously produced dissolved nutrients coming from upstream [72] and/or possibly from marsh outwelling. Chanton and Lewis [72] also found that although consumers primarily utilize autochthonously produced substrates during periods of high river flow, the influx of terrestrial floodplain detritus does augment productivity in the bay. They concluded that reduced river flow would have a detrimental effect on overall estuarine production, especially during seasonal and extended droughts. The trophic status of the bay is therefore intricately linked to Apalachicola River's hydrologic regime, which impacts the bay.

As stated earlier, Deepwater Horizon Oil Spill has not significantly impacted Apalachicola Bay oysters. Compared to other areas along the Gulf Coast, the water quality in Apalachicola

Bay has remained relatively unaffected, with the exception of tar balls washing ashore and some oil sheens. To enhance oyster landings after the spill, the state of Florida opened a 7-day-per-week oyster harvest at both the summer and winter harvesting grounds during June–August 2010; however, preliminary commercial landings reports suggest that Franklin County's 2010 oyster harvest was the lowest in 5 years, although landings from 2007 and 2009 were the highest in 20 years. Furthermore, the annual 2010 landings rate (pounds/trip) was the lowest since 1991 (**Figure 13**). This drop in oyster production and the over-tapping of the winter oyster beds translated into lost revenue, which reverberated throughout the Florida seafood and restaurant. The oyster fishery has not rebounded since 2011 and prices have increased (**Figure 13**). University of Florida researchers developed a population model to determine whether harvest of sub-legal oysters contributed to the crash in the Apalachicola Bay fishery while they found an increase in natural mortality [39, 78].

Since natural oyster populations have been unable to keep up with demand let alone sustainable historic stocks, aquaculture has become an essential part of restoration and stock enhancement efforts. According to the California Aquaculture Association [79], the top U.S. marine aquaculture species was oysters. In 2015, U.S. shellfish farmers produced 15,876 metric tons of oysters at $173 million market value [80]. According to Stewart [81], the farmed oyster production grew by 806% between 2006 and 2012 in Chesapeake Bay. Rheault (Executive Director of East Coast Shellfish Association) in Stewart's report stated that the east coast shellfish production for oysters has doubled in 5 years at a steady rate of 12% per year. With $2 a pop on U.S. restaurant menus, associated demand for oysters are making oyster farming vital player in the United States.

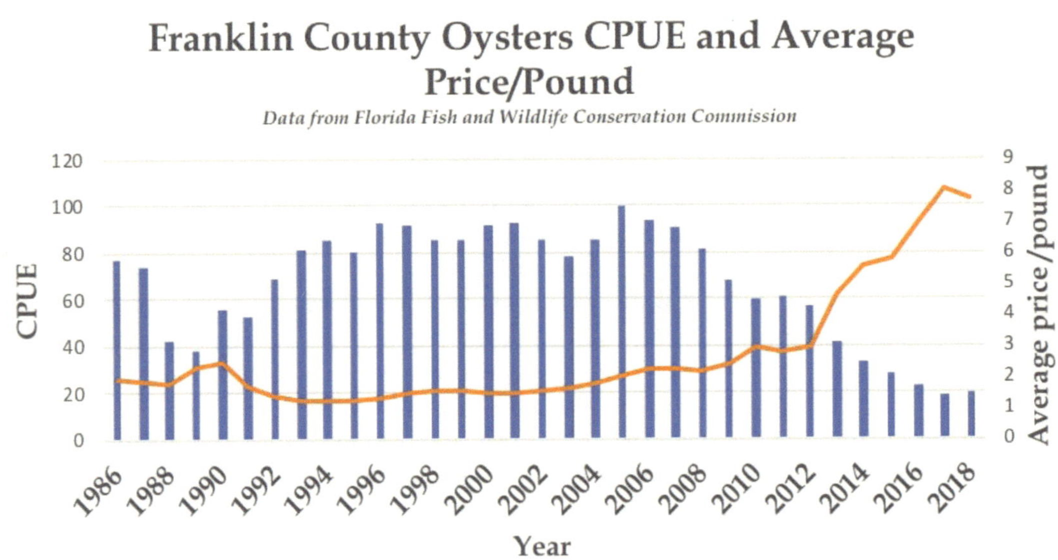

Figure 13. Oysters catch per unit effort and average price per pound of oysters (*Data from Florida Fish and Wildlife Conservation Commission http://myfwc.com/, Figure by Smith (unpublished)*).

5. Conclusion

Here we discussed two different estuary systems with differences in ecosystem management goals and plans with different ecosystem indicators (TSS and turbidity in Delaware Inland Bays versus salinity in Apalachicola Bay), however both estuaries have few common issues mainly due to increase population and population driven activities (*see* **Table 3**). These issues include frequent eutrophication events, increase pollutants via storm water runoff, agricultural or residential areas, overfishing and habitat alterations. Both estuary systems have some indicators different from each other such as coral reefs and essential fish habitat for Apalachicola Bay versus essential fish habitat or areas that will be open to shellfish harvesting for the Delaware Inland Bays. Major alterations and changes to those fragile ecosystems are mainly due to anthropogenic activities and management goals for both estuaries should be to minimize further changes and mitigate areas already altered.

Although decision and management strategies will be different, each estuary system or watershed, depending on the critical areas of concerns and related activities, solution is dependent upon how we plan our next action. Are we setting our goal for too short term or

Criteria	Delaware Inland Bays, DE	Apalachicola Bay, FL
Characteristics	• Three low flushing interconnected bodies of water (Indian River Bay, Little Assawoman Bay, and Rehoboth Bay). • Watershed - 811 km^2 in Delaware • Average water depth of 1.2 meters • Frequent alteration to waterways?	• River-dominated estuary and lagoon in Florida. • Watershed - 540 km^2 in Alabama, Georgia and Florida • Average water depth of 2 meters • Pristine system?
Oyster Population Status	• Very limited oyster population • Restoration and mitigation are necessary	• Natural oyster population dwindling, not rebounding
Aquaculture Status	• Aquaculture permits have been issued and approval obtained • However, implementation is very slow	• Natural oyster population is declining • Aquaculture is becoming popular
Challenges	• Frequent eutrophication • Increased pollutants from agricultural and residential runoff • Overfishing • Habitat alterations • Increase land uses due to human population and related activities • Human interference of habitat • Very low natural oyster population and recruitment and approval and implementation of aquaculture	• Reduced freshwater input due to drought and upriver usage • Estuary salinity increase • Increased predation • Most common issues are decrease in oyster population, approval of oyster aquaculture

Criteria	Delaware Inland Bays, DE	Apalachicola Bay, FL
Management Practices	• Short-term management practices primarily on nutrient reduction • Many users of the watershed requiring frequent accommodations • Frequent conflict resolution delaying implementation plans	• Long-term management plan focusing on limiting harvesting
Opportunities	• Good oyster growth and survival at designated oyster aquaculture sites are expected. • Potential for clam aquaculture	• Increased oyster culture • Potential for mussel aquaculture

Table 3. Comparison of two estuaries with differences in water quality, anthropogenic impacts, and management goals.

are we having a comprehensive plan? Either it is re-introduced to the area in the case of the Delaware Inland Bays or naturally occurring in Apalachicola Bay, oysters provide ecosystem services long proven and sustainability of these ecosystems lies on the comprehensive and integrated ecosystem planning and assessment. Although integrated ecosystem assessment plan is not available for the Delaware Inland Bays with promising nutrient reduction and waterway improvement initiatives with leadership of the Center for Inland Bays, there are few applied for Apalachicola Bay that provides foundation for assessing the merging needs of the area from the ecosystem health perspectives?

Whether aquaculture is used for revitalizing habitat or restoring native species or human consumption, there are big variation the way each operate. Growing demand for fresh seafood has prompted a long-term viable and sustainable aquaculture industry worldwide. With wild capture fisheries exceeding the maximum sustainable harvest capacity, aquaculture has become a bridge in closing the gap between rising demand and seafood sources. By 2011, farmed seafood accounts over 50% of overall production in the global marketplace [82]. As stated clearly by Shumway et al. [83] "Shellfish are one of the best candidates for ecologically sustainable aquaculture. Farming of shellfish not only provides a high quality, high value, sustainable harvest from the ocean, it also provides jobs and social and economic development, all while providing tangible benefits to the marine environment. A productive shellfish farm means a healthy and equally productive surrounding environment let's give the lowly molluscs their due!"

Restoring oyster population requires further elevation aquaculture has and will enhance the ecosystem health of both watershed discussed in this chapter. Either, we cease wild harvesting or we provide the push to enhance the population and in case of oysters, we are hopeful "aquaculture" will provide both environmental and economical stability in those bays.

Acknowledgements

Projects discussed in this chapter are funded by USDA Evans-Allen Grants DELXDSUGO2015, 2016- DELX12345, USDA-NIFA 2013-38821-21246, NSF-EPSCOR Grant Number IIA-1301765, and NOAA-ECSC Grant Number NA11SEC4810001.

http://dx.doi.org/10.5772/intechopen.78989

Author details

Gulnihal Ozbay* and Stacy L. Smith

*Address all correspondence to: gozbay@desu.edu

Department Agriculture and Natural Resources, Delaware State University, Dover, DE, USA

References

[1] Chaillou JC, Weisberg SB, Kutz FW, DeMoss TE, Mangiaracina L, Magnien R, Eskin R, Maxted J, Price K, Summers JK. Assessment of the Ecological Condition of the Delaware and Maryland Coastal Bays. Annapolis, Maryland, USA: U.S. Environmental Protection Agency, Region III; 1994. p. 119

[2] Zimmerman RJ, Minello TJ, Baumer MC, Castiglione M. Oyster Reef as Habitat for Estuarine Macrofauna. Galveston, TX: NOAA Technical Memorandum, NMFS-SEFC-249; 1989. p. 16

[3] Newell RIE. Ecosystem influences of natural and cultivated populations of suspension-feeding bivalve mollusks: A review. Journal of Shellfish Research. 2004;**23**:51-61

[4] Takacs R, King J, Jasinski P. Oyster restoration on the east and gulf coasts. NOAA Habitat Connections. 2005;**5**(2):1-4

[5] Brumbaugh RD, Sorabella LA, Oliveras Garcia C, Goldsborough WJ, Wesson JA. Making a case for community-based oyster restoration: An example from Hampton roads, Virginia, USA. Journal of Shellfish Research. 2000;**19**:467-472

[6] Kennedy VS. The ecological role of the eastern oyster, *Crassostrea virginica*, with remarks on disease. Journal of Shellfish Research. 1996;**15**:177-183

[7] Rothschild BJ, Ault JS, Goulletquer P, Heral M. Decline of the Chesapeake Bay oyster population: A century of habitat destruction and overfishing. Marine Ecology Progress Series. 1994;**111**:29-39

[8] Marenghi FP. Floating oyster (*Crassostrea virginica*) aquaculture as habitat for fishes and macroinvertebrates in Delaware's Inland Bays. Master's Thesis in Natural Resources. Delaware State University: Dover, Delaware, USA; 2009. 203 pp

[9] Goldsborough W, Meritt D. Oyster Gardening for Restoration and Education. Maryland Sea Grant College Program. College Park, Maryland: Oyster Alliance; 2001. p. 12

[10] Luckenbach MW, O'Beirn FX, Taylor J. An Introduction to Culturing Oysters in Virginia. Gloucester Point, Virginia: School of Marine Science Virginia Institute of Marine Science College of William and Mary; 1999. p. 24

[11] Delaware Inland Bays Estuary Program Report (DIBEP). Science and Technical Advisory Committee, Delaware Inland Bays Estuary Program Characterization Summary. Dover, Delaware. 1993

[12] Delaware Center for the Inland Bays Report. Oyster Gardening Program. 2008. http://darc.cms.udel/ibog/index.html [Accessed: April 12, 2017]

[13] Delaware Center for the Inland Bays Report. The Delaware Center for Inland Bays. Report on State of the Delaware Inland Bays. Final Report of the Shellfish Aquaculture Tiger Team to the Board of Directors. 2011. http://www.inlandbays.org/wp-content/documents/2011-state-of-the-bays.pdf [Accessed: April 12, 2015]

[14] Walch M, Seldomridge E, McGowan A, Boswell S, Bason C. 2016 State of the Delaware Inland Bays. Rehoboth Beach, Delaware: The Delaware Center for the Inland Bays, Rehoboth Beach Delaware; 2016. p. 72

[15] Martin DM, Morton T, Dobrzynski T, Valentine B. Estuaries on the Edge: The Vital Link between Land and Sea. Washington, DC: The American Oceans Campaign; 1996. p. 297

[16] U.S. Environmental Protection Agency. 2011. State of the Delaware Inland Bays. https://www.epa.gov/sites/production/files/2015-09/documents/2011-state-of-the-bays.pdf [Accessed: January 17, 2017]

[17] Grabowski JH, Conrad RF, Keeler AG, Opaluch JJ, Peterson CH, Piehler MF, Powers SP, Smyth AR. Economic valuation of ecosystem services provided by oyster reefs. Bioscience. 2012;**62**(10):900-909

[18] Erbland P. A Comparison of the Macrofaunal Communities Inhabiting a Created Oyster (*Crassostrea virginica*) reef and oyster aquaculture gear. Master's Thesis. Delaware State University: Dover, Delaware. 2007. 87 pp

[19] Erbland PJ, Ozbay G. A comparison of the macrofaunal communities inhabiting a *Crassostrea virginica* oyster reef and oyster aquaculture gear in Indian River Bay, Delaware. Journal of Shellfish Research. 2008;**27**:757-768

[20] Marenghi FP, Ozbay G. Floating oyster, *Crassostrea virginica* Gmelin 1791, aquaculture as habitat for fishes and macro-invertebrates in Delaware inland bays: The comparative value of oyster clusters and loose Shell. Journal of Shellfish Research. 2010a;**29**(4):889-904

[21] Marenghi FP, Ozbay G. Preliminary habitat assessment of floating oyster (*Crassostrea virginica*) gardens (Delaware). Restoration Ecology Journal. 2010b;**28**(3):254-257. DOI: 10.3368/er.28.3.254

[22] Marenghi F, Ozbay G, Rossi-Snook K, Chalabala EJ. Restoration programs in inland bays improves ecosystems as oyster populations recover. Global Aquaculture Alliance Global Aquaculture Advocate. 2009:16-17

[23] Reckenbeil BA. Assessment of oyster restoration along human altered shorelines in the Delaware Inland Bays: An examination of riprap stocked with the eastern oyster (*Crassostrea virginica*). Master's Thesis in Natural Resources. Delaware State University: Dover, Delaware, USA; 2013. 189 pp

[24] Rossi-Snook K, Ozbay G, Marenghi F. Oyster (*Crassostrea virginica*) gardening for restoration in Delaware Inland Bays, USA. Aquaculture International Journal. 2009;**18**(1): 61-67. DOI: 10.1007/s10499-009-9271-5

[25] Kee E. Delaware Agriculture. 2010. http://www.nass.usda.gov/Statistics_by_State/Delaware/Publications/DE%20Ag%20Brochure_web.pdf [Accessed: January 12, 2015]

[26] Delaware Center for the Inland Bays Report. The Power of People and Nature to Restore the Waters of the Inland Bays. 2015. 8 pp. https://www.inlandbays.org/wp-content/uploads/2-2-2016-CIB-Annual-Report-2015-FINAL.pdf [Accessed: April 22, 2017]

[27] Delaware Center for the Inland Bays Report. New Faces Changing Landscapes. The Delaware Center for Inland Bays: Issues Affecting the Bays. 2014. 8 pp. http://www.inlandbays.orgabout-the-bays [Accessed: April 12, 2015]

[28] Committee on Environment and Natural Resources. 2010. Scientific Assessment of Hypoxia in U.S. Coastal Waters. Interagency Working Group on Harmful Algal Blooms, Hypoxia, and Human Health of the Joint Subcommittee on Ocean Science and Technology: Washington, DC, 164 pp

[29] Kellogg ML, Cornwell JF, Owens MS, Paynter KT. Denitrification and nutrient assimilation on a restored oyster reef. Marine Ecology Progress Series. 2013;**480**:1-19. DOI: 10.3354/meps10331

[30] Rose JM, Bricker SB, Ferreira JG. Comparative analysis of modeled nitrogen removal by shellfish farms. Marine Pollution Bulletin. 2015;**91**(1):185-190. DOI: 10.1016/j.marpolbul.2014.12.006

[31] DNREC Division of Watershed Stewardship. Shellfish Growing Waters. 2015. http://www.dnrec.delaware.gov/swc/wa/Pages/Shellfish-Growing-Waters.aspx [Accessed: January 12, 2015]

[32] DNREC Division of Fish and Wildlife. Proposed Shellfish Aquaculture Development Areas. 2014. http://www.dnrec.delaware.gov/fw/fisheries/pages/shellfishaquaculture.aspx [Accessed: January 12, 2015]

[33] Livingston RJ. The Ecology of the Apalachicola Bay System: An Estuarine Profile. U.S. Fish and Wildlife Service FW/OBS 82/05; 1984, 148 pp

[34] McNulty JK, Lindall Jr WN, Sykes JE. Cooperative Gulf of Mexico Estuarine Inventory and Study, Florida: Phase I, Area Description. OAA Tech Report. NMFS Circ-368; 1972. 126 pp

[35] Iseri KT, Langbein WB. Large rivers of the United States: U.S. Geological Survey Circular 686; 1974, 10 pp

[36] Light HM, Vincent KR, Darst MR, Price FD. Water-Level Decline in the Apalachicola River, Florida, from 1954 to 2004, and Effects on Floodplain Habitats: U.S. Geological Survey Scientific Investigations Report 2006-5173; 2006. 83 pp

[37] USGS. Coastal and Marine Geology Program. U.S. Geological Survey Open File Report 2006-1381. Geophysical Mapping of Oyster Habitats in a Shallow Estuary; Apalachicola Bay, Florida. 2006. https://cmgds.marine.usgs.gov/publications/of2006-1381/html/maps.htm [Accessed: April 30, 2018]

[38] ***U.S. Department of the Interior, U.S. Geological Survey. USGS current conditions for the nation. USGS 02359170 Apalachicola River NR Sumatra, FL. USGS Sumatra gage. 2018. Website https://waterdata.usgs.gov/nwis/uv?02359170 [Accessed: April 30, 2018]

[39] Florida Sea Grant and University of Florida and authors therein. Apalachicola Bay Oyster Situation Report. 2013. TP-200

[40] NWF National Wildlife Federation (NWF)/Florida Wildlife Federation (FWF). An Unfavorable Tide Global Warming, Coastal Habitats and Sportfishing in Florida; 2006. 60 pp

[41] Havens K, Allen M, Irani T, Lindsey A, Morris JG, Kane A, Kimbro D, Otwell S, Pine B, Walters C. Apalachicola Bay Oyster Situation Report. Florida Sea Grant TP-200: University of Florida; 2013. p. 32

[42] Jones CG, Lawton JH, Shachak M. Organisms as ecosystem engineers. Oikos. 1994;**69**: 373-386

[43] Korringa P. Recent advances in oyster biology. The Quarterly Review of Biology. 1952; **27**(266-308):339-365

[44] Wells HW. The Fauna of oyster beds, with special reference to the salinity factor. Ecological Monographs. 1961;**31**(3):239-266. DOI: 10.2307/1948554. http://www.jstor.org/stable/1948554

[45] Newell RIE. Ecological changes in Chesapeake Bay: Are they the result of overharvesting the American oyster, *Crassostrea virginica?* In: Lynch MP, Krome EC, editors. Understanding the Estuary: Advances in Chesapeake Bay Research. Gloucester Point, Virginia: Chesapeake Research Consortium, Publication 129 CBP/TRS 24/88; 1988. pp. 536-546

[46] Bahr LM, Lanier WP. The Ecology of Intertidal Oyster Reefs of the South Atlantic Coast: A Community Profile. Washington, D.C. FWS/OBS-81/15: U.S. Fish and Wildlife Service, Office of Biological Services; 1981. p. 105

[47] Breitburg DL. Are three-dimensional structure and healthy oyster populations the keys to an ecologically interesting and important fish community? In: Luckenbach MW, Mann R, Wesson JA, editors. Oyster Reef Habitat Restoration: A Synopsis and Synthesis of Approaches. Gloucester Point, Virginia: VIMS Press; 1998. pp. 239-250

[48] Posey MH, Alphin TD, Powell CM, Townsend E. Oyster reefs as habitat for fish and decapods. In: Luckenbach MW, Mann R, Wesson JA, editors. Oyster Reef Habitat Restoration: A Synopsis of Approaches. Williamsburg, Virginia: Virginia Institute of Marine Science Press; 1999. pp. 229-237

[49] Rodney WS, Paynter KT. Comparisons of macrofaunal assemblages on restored and non-restored oyster reefs in mesohaline regions of Chesapeake Bay in Maryland. Journal of Experimental Marine Biology and Ecology. 2006;**335**:39-51

[50] Piazza BP, Banks PD, LaPeyre MK. The potential for created oyster shell reefs as a sustainable shoreline protection strategy in Louisiana. Restoration Ecology. 2005;**13**:499-506

[51] Kitchell JE, O'Neill RV, Webb D, Gallepp G, Bartell SM, Koonce JF, Ausmus BS. Consumer regulation of nutrient cycling. Bioscience. 1979;**29**:28-34

[52] Malone TC, Kemp WM, Ducklow HW, Boynton WR, Tuttle JH, Jonas RB. Lateral variation in the production and fate of phytoplankton in a partially stratified estuary. Marine Ecology Progress Series. 1986;**32**:149-160

[53] Newell RIE, Fisher TR, Holyoke RR, Cornwell JC. Influence of eastern oysters on nitrogen and phosphorus regeneration in Chesapeake Bay, USA. In: Dame R, Olenin S, editors. The Comparative Roles of Suspension Feeders in Ecosystems. NATO Science Series IV Earth and Environmental Sciences. Netherlands: Springer; 2005. pp. 93-120

[54] Newell RIE, Cornwell JC, Owens MS. Influence of simulated bivalve biodeposition and microphytobenthos on sediment nitrogen dynamics, a laboratory study. Limnology and Oceanography. 2002;**47**:1367-1379

[55] Boucher G, Boucher-Rodoni R. In situ measurement of respiratory metabolism and nitrogen fluxes at the interface of oyster beds. Marine Ecology Progress Series. 1988;**44**:229-238. DOI: 10.3354/meps044229

[56] Dame RF. Oyster refs as complex systems. In: Dame RF, Olenin S, editors. The Comparative Roles of Suspension-Feeders in Ecosystems. Dordrecht, The Netherlands: Kluwer Academic Publishers; 2005. pp. 331-344

[57] Prins TC, Smaal AC, Dame RF. A review of the feedbacks between bivalve grazing and ecosystem processes. Aquatic Ecology. 1998;**31**(4):349-359

[58] Jackson AR, Rouse DR, Ozbay G. The effects of oyster (*Crassostrea virginica*) structure and bioactivity on water quality, sediment nutrients, Infaunal communities, and transient fish biomass. Journal of Environmental Monitoring and Restoration. 2008;**4**:66-82

[59] Fuoco M. Nitrogen source tracking and benthic community assessment in relation to oyster aquaculture in the Delaware inland bays. Master's Thesis in Natural Resources. Delaware State University: Dover, Delaware, USA; 2018. 71 pp

[60] Ulanowicz RE, Tuttle JH. The trophic consequences of oyster stock rehabilitation in Chesapeake Bay. Estuaries. 1992;**15**:298-306

[61] Marenghi F, Ozbay G, Erbland P, Rossi-Snook K. A comparison of the habitat value of sub-tidal and floating oyster (*Crassostrea virginica*) aquaculture gear with a created reef in Delaware's inland bays, USA. Aquaculture International Journal. 2009b;**18**:69-81. DOI: 10.1007/s10499-009-9273-3

[62] Ozbay G, Reckenbeil B, Marenghi F, Erbland P. Eastern oyster (*Crassostrea virginica*) aquaculture and diversity of associated species. In: Turner JP, editor. Oysters: Biology, Consumption, and Ecological Importance. NOVA Publishing. Book Chapter; 2013. p. 128. ISBN: 978-1-62948-808-0

[63] Ozbay G, Marenghi F, Rossi-Snook K, Ewart J, Chalabala EJ. 2009. Delaware Oyster Gardening and Restoration - A Cooperative Effort. YSI Application Note. YSI Aquaculture 0409 A571. 2009 YSI Inc. 3 pp

[64] Delaware Center for the Inland Bays Report. Oyster (*Crassostrea virginica*) Gardening Program for Restoration in Delaware's Inland Bays, USA. 2013. https://www.researchgate.

net/publication/226882567_Oyster_Crassostrea_virginica_gardening_program_for_restoration_in_Delaware's_Inland_Bays_USA [Accessed: April 22, 2017]

[65] Fulford RS, Breitburg DL, Newell RIE, Kemp WM, Luckenbach M. Effects of oyster population restoration strategies on phytoplankton biomass in Chesapeake Bay: A flexible modeling approach. Marine Ecology Progress Series. 2007;**336**:43-61. DOI: 10. 3354/meps336043

[66] Pomeroy LR, D'Elia CF, Schaffner LC. Limits to top-down control of phytoplankton by oysters in Chesapeake Bay. Marine Ecology Progress Series. 2006;**325**:301-309. DOI: 10.3354/meps325301

[67] Buitrago E, Alvarado D. A highly efficient oyster spat collector made with recycled materials. Aquaculture Engineering. 2005;**33**:63-72

[68] Ozbay G, Cannon A. Gauging benefits of oyster garden restoration in Delaware. World Aquaculture Society. 2017;**48**(4):59-61

[69] Edmiston HL. A River Meets the Bay: A Characterization of the Apalachicola River and Bay system. Apalachicola National Estuarine Research Reserve (ANERR). ANERR National Estuarine Research Reserve 2014. 2008. http://www.nerrs.noaa.gov/Apalachicola/welcome.html [Accessed: April 12, 2018]

[70] Galtsoff PS. The American oyster *Crassostrea virginica* Gmelin. Fisheries Bulletin. 1964;**64**: 1-480

[71] FDEP. Apalachicola-Chattahoochee-Flint River System. 2017. http://www.dep.state.fl.us/mainpage/acf/timeline.htm [Accessed: January 12, 2015]

[72] Chanton J, Lewis FG. Examination of coupling between primary and secondary production in a river-dominated estuary: Apalachicola Bay, Florida, U. S. A. Limnology and Oceanography. 2002;**47**:683-697

[73] Livingston RJ, Lewis III FG, Woodsum GC, Niu X, Howell IV RL, Ray GL. Modeling oyster population response to variation in freshwater input. Estuarine, Coastal and Shelf Science. 2000;**50**:655-672

[74] Livingston RJ, Iverson RL, Estabrook RH, Keys VE, Taylor Jr J. Major features of the Apala chicola Bay system: Physiography, biota, and resource management. Florida Scientist. 1974;**4**:245-271

[75] Childs CR, Rabalais NN, Turner RE, Proctor LM. Erratum: Sediment denitrification in the Gulf of Mexico zone of hypoxia: Marine Ecology Progress Series 247; Cambria. pp. 310-310

[76] Mortazavi B, Iverson RL, Landing WM, Lewis FG, Huang WR. Control of phytoplankton production and biomass in a river-dominated estuary: Apalachicola Bay, Florida, USA. Marine Ecology Progress Series. 2000a;**198**:19-31

[77] Mortazavi B, Iverson RL, Huang WR, Lewis FG, Caffrey J. Nitrogen budget of Apalachicola Bay, a bar-built estuary in the northeastern Gulf of Mexico. Marine Ecological Progress Series. 2000b;**195**:1-14

[78] Alderstein D. State to Expand Aquaculture Leases. The Apalachicola and Carrabelle Times. 2018. Online. http://www.apalachtimes.com/news/20180405/state-to-expand-aquaculture-leases [Accessed: on April 22, 2018]

[79] California Aquaculture Association. NOAA Releases Fisheries of the United States, 2016 Report. 2017. October 30, 2017. Chico, CA 95928 info@caaquaculture.org. http://caaquaculture.org/2017/10/30/noaa-releases-fisheries-of-the-united-states-2016-report/[Accessed: April 26, 2018]

[80] NOAA Fisheries-Office of Aquaculture. Shellfish Aquaculture: How Much Is U.S. Shellfish Aquaculture Worth? Report by December 4, 2017. 2017. https://www.fisheries.noaa.gov/insight/shellfish-aquaculture. [Accessed: May 18, 2018]

[81] Stewart J. Entrepreneur-driven US oyster industry growing on farm-to-table movement. Undercurrent News, January 27, 2014. 2014. https://www.undercurrentnews.com/2014/01/27/entrepreneur-driven-us-oyster-industry-growing-on-farm-to-table-movement/ [Accessed: May 20, 2018]

[82] Bush SR, Toonen H, Oosterveer P, Mol AP. The 'devils triangle' of MSC certification: Balancing credibility, accessibility and continuous improvement. Marine Policy. 2013;**37**: 288-293. DOI: 10.1016/j.marpol.2012.05.011

[83] Shumway SE, Davis C, Downey R, Karney R, Kraeuter JParsons J, Rheault R, Wikfors G. Shellfish aquaculture: In praise of sustainable economies and environments. World Aquaculture. 2003;**34**:15-18

[84] Ozbay G, Phalen L. A Comparison of Bottom, Hanging, and Floating Gear for Growing Eastern Oysters (Crassostrea virginica) in Broadkill River. Delaware Oyster Aquaculture Technical Paper 1. Dover, DE: Delaware State University; 2017. pp. 6

[85] NOAA declares fishery disaster for Apalachicola oysters. http://blog.al.com/gulf-coast/2013/08/noaa_declares_fishery_disaster.html [Accessed on June 25, 2018]

[86] Tagliapietra D, Sigovini M. Benthic fauna: Collection and identification of macrobenthic invertebrates. NEAR Curriculum in Natura Environmental Science, Terre et Environnement. 2010;**88**:253-261. ISBN: 2-940153-87-6

Section 3

Homeopathy and Probiotics

Chapter 4

Aquacultural Homoeopathy: A Focus on Marine Species

Mazón-Suástegui José Manuel,
Tovar-Ramírez Dariel, Salas-Leiva Joan Sebastian,
Arcos-Ortega Guadalupe Fabiola,
García-Bernal Milagro,
Avilés-Quevedo María Araceli,
López-Carvallo Jesús Antonio,
García-Corona José Luis,
Ibarra-García Laura Elizabeth,
Ortíz-Cornejo Nadia Livia, Teles Andressa,
Rosero-García Antonia, Abasolo-Pacheco Fernando,
Campa-Córdova Ángel Isidro, Saucedo Pedro,
Barajas-Frías José Delfino, Ormart-Castro Pablo,
Rodríguez-Jaramillo Carmen,
González-González Reinaldo, Barajas-Ponce Ulysses,
Tordecillas-Guillén Jorge Luis,
Álvarez-Gil Francisco Antonio,
Pineda-Mahr Gustavo, Peiro-López Jesús and
Robles-Mungaray Miguel

Additional information is available at the end of the chapter

http://dx.doi.org/10.5772/intechopen.78030

Abstract

Homoeopathy is an alternative medical system proposed by Samuel Hahnemann in the eighteenth century. It uses highly diluted and agitated substances that derived from plants, minerals or animals, which have shown to be effective in human medicine, agronomy, veterinary, and as a novelty, in marine aquaculture. Aquacultural homoeopathy has developed rapidly in recent years, partially motivated by the misuse of powerful drugs (hormones, antibiotics, disinfectants) that when solving a problem generate undesirable side effects. In the last 10 years, scientific articles have been published on its application in freshwater fish native to Brazil, obtaining beneficial effects on growth, survival, hepatosomatic index, development of muscle fibres and lipid content in muscle. At Centro de Investigaciones Biológicas del Noroeste (CIBNOR, Mexico: www.cibnor.mx), we have studied the effects of homoeopathy to improve the culture of economically important marine species of molluscs, fish and shrimp. In this chapter, we show a selection of different research with preliminary or advanced results, related to the use of homoeopathy and its impact on zootechnic, biochemical, genomic and transcriptomic parameters in marine molluscs, fish and crustaceans. The results obtained suggest that homoeopathy is an eco-friendly alternative applicable in aquaculture industry to improve various productive and health aspects.

Keywords: homeopathic medicine, marine molluscs, crustaceans and fish, growth, survival and biomass production, physiology and reproduction, genomic, metagenomic and transcriptomic response

1. Introduction

Aquaculture has become a dynamic industry in constant development with the fastest growth food production sector of animal origin, even at a higher rate than human population, which could provide half of the fish consumption worldwide. However, its great development has turned out into great problems and challenges, based mainly on hyper-intensive production systems where the common denominator is high cultivation densities that generally cause physiological stress. Acute or chronic stress acts in a synergistic way to other environmental factors, negatively impacting productive parameters or increasing sensitivity to diseases produced by opportunistic pathogens as bacteria and viruses. Altogether finally translates into great economic losses to producers. Among the solutions for these problems is the use of expensive immunostimulant substances, so producers have opted to use and misuse antibiotics and/or prophylactics to treat stress and its consequences. Antibiotics promote the resistance of target organisms and leave residues in the environment and in tissues, which has become a public health problem in some countries [1, 2]. For this reason, alternatives have been sought such as the use of probiotics [3], phytobiotics [4], and recently homoeopathy that has been investigating as a novel alternative to improve various productive aspects in the culture of aquatic organisms [5]. Homoeopathy is a branch of universal medicine based on the Law of Similars also expressed as "principle of like" (*Similia Similibus Curentur* = Like cures Like), and it is applied in ultra-diluted and successed minimal doses. It is assumed that a substance that is applied in high doses (massively) generates a pathological symptomatology, so if it is applied in minimum doses (obtained by serial dilution and agitation), it can in turn cure it [6]. Homoeopathy derives from a Hippocratic medical concept proposed by the German physician Samuel Hahneman (1755–1843), who developed decimal (D; 1: 9), centesimal (C; 1: 99) millesimal (M; 1: 999) dilutions and other medicines, which are actually known as homoeopathic "dilutions" or "dynamisations." This process consists of serial dilution of mineral, plant and animal concentrate materials in water-ethanol vehicle, and vigorous agitation or "succussion" [7]. The starting point to obtain a certain dynamisation is a concentrate or Mother Tincture (MT), which is an alcoholic extract prepared from plants, animals, minerals, and even nanoparticulate metals [7]. In spite of its high dilution, it is possible to detect nanoparticles of the "ponderable active principle" (MT) in the dynamisations, even in high centesimal dilutions despite the fact that according to Avogadro's theory, they should not have a single molecule of MT [8]. Therefore, re-naming homoeopathy as "Adaptative Network Nanomedicine" has been recently proposed [9].

Homoeopathy has the peculiarity of stimulating the self-recovery of dynamic homeostasis when it has been lost due to exogenous and/or endogenous factors. Therefore, it does not focus on "the disease" but on the manifestations and intrinsic defence mechanisms of the "patient." It offers "signals" of systemic action to the treated individual to promote its self-regulation to recover homeostasis, and as it uses ultra-diluted minimal doses, it does not leave residues in the organism or in the environment. As a counterpart, the other medicine known as "allopathy" derives from a galenic concept based on the "principle of opposites" and the application of massive doses of various chemotherapeutic agents officially classified as anti-microbial, anti-viral, anti-inflammatory, anti-spasmodic, anti-histaminic, anti-fever and other "anti" drugs. Homoeopathic medicines can be administered to any living being, including terrestrial and aquatic plants, wild animals in captivity and breeding, as well as freshwater and marine species of commercial interest [7]. It has been widely used in human, animal and plant medicine because it induces specific responses and increases immunity, favouring resistance to pathogens under stressing situations, promoting a better post-infection recovery and improving internal dynamic homeostasis [5, 6, 10].

In Mexico, homoeopathy is recognised as a therapy of alternative medicine, and its practice was authorised by presidential decree in July 31, 1895; today its study and practice is officially recognised in the general health law (2015), and only health professionals can prescribe homoeopathic medicines, which must have an official code. This is the way homoeopathic medicines are differentiated from products like herbs for infusion and herbal remedies. In countries such as Brazil, there are homoeopathic medicines exclusively for veterinary use, for marine and freshwater fishes, registered with the Ministry of Agriculture. Important and promising results have been reported in freshwater organisms, mainly Nile Tilapia *Oreochromis niloticus* and Pacu *Piaractus mesopotamicus* [11–13]. Taking into account these antecedents, this chapter compiles not only the experimental results obtained in marine organisms, such as molluscs, fish and crustaceans when treated with commercial homoeopathic medicines for human use and approved by Federal Health Law and Health Ministry of México but also other ones that have been designed and developed at Centro de Investigaciones Biológicas del Noroeste (CIBNOR) in La Paz, Baja California Sur, Mexico.

For the purposes of this chapter, commercial drugs for human use have been utilised in the form of liquid hydro-alcoholic dynamisations (Similia® Laboratories, Mexico), injectable aqueous dynamisation (Rubiopharma®, Mexico) or sugar impregnated with homoeopathic complexes (Arenales Homoeopathy®, Brazil). They were considered "stock dynamisations," from which the respective "work dynamisations" were obtained through a serial process of dilution-succussion decimal or centesimal. Other non-commercial medicines were designed and developed at CIBNOR, from bacterial products (nosodes). This chapter deals with unitary or complex laboratory treatments used and registered with the Mexican Ministry of Health, such as Similia® and Rubiopharma® and a Cuban drug of Labiofam®, which will be described as follows: *Passiflora incarnata, Valeriana officinalis, Ignatia* amara and *Zincum valerianicum* of

Similia® (PaV); Cyme-Heel, Gal-Heel, Hepa-Heel, Mucs-Heel and Chol-Heel of Rubiopharma® (INM); Endecto (END) and Infecçoes (INF) of Arenales Homoeopathy®; *Phosphoric acid* (PhA), *Phosphoric acid* (AcF), *Silicea terra* (SiT), *Sodium methasilicate* (MsS), Scorpion toxin Vidatox® (ViT); *Calcium sulphuricum* (CaS); *Hepar sulphuris* (HeS), *Ferrum phosphoricum* (FeP); *Zincum phosphoricum* (ZiP); *Magnesium phosphoricum* (MaP), *Mercurius solubilis* (MeS). Also, other nosode-type homoeopathic medicines from *Vibrio* compounds (ViP, ViA) were applied. These and other nosode type HOM-products have been designed by CIBNOR that is processing the respective trademark and industrial property titles (Office for industrial protection and technology transfer; OTT-CEPAT/CIBNOR; www.cibor.gob.mx). Ethanol (ET) and no-HOM nor ethanol (NT) were used as control treatments. Homoeopathic medicines (HOM) were sprinkled on balanced food or inert sugar pills or added directly to culture seawater.

2. Effects of homoeopathy in marine species

2.1. Mollusc

2.1.1. Catarina scallop (Argopecten ventricosus)

The production of mollusc bivalves around the world is still challenged every year by the propagation and emergence of new diseases. Scallops are especially susceptible to epizootic pathogenic bacteria in the hatchery, mainly those related to *Vibrio* spp. and *Aeromonas* spp. [14]. Four experiments described below were performed with *A. ventricosus*.

Trial 1. To assess the effects of homoeopathic treatments (HOM treatments) in the immune system of Catarina scallop *A. ventricosus*, an experimental design was applied at CIBNOR during nursery management with five homoeopathic treatments and three controls, four replicates each. Juveniles were placed in recirculating upwelling nursery units (40 l), each one with four PVC upwelling cylinders for 21 days. Continuous aeration and a microalgal food mix 1:1 (*Isochrysis galbana - Chetoceros calcitrans*; 150,000 cel ml^{-1}) were provided. The following HOM treatments and controls were applied: ViP (T1), ViA (T2), PhA Metasilicate (T3), PhA-SiT (T4), ViT (T5), dynamised ET (T6), diluted ET (T7) and NT nor ethanol added (T8).

At the end of the assay, mainly nosodes (ViP, ViA) formulated from pathogenic bacterial compounds, stimulated a significant ($p < 0.05$) growth rate and increased haemocyte count which were counted using three images at 100X from scallops tissue were processed by Image Pro Plus 6.0 to count the number of haemocytes in a tissue area (0.21 mm^2). Haemocytes count was 1–3 times higher than controls (**Table 1**). These results suggested immune system enhancement by the action of HOM treatments because it is known they stimulate enzymatic activity of superoxide dismutase (SOD) and catalase (CAT) related with antioxidant responses of the organisms during oxidative stress [15]. The proliferation of haemocytes is related to activation of immune response because they are the primary cells responsible to protect organisms against infections [16]. Antioxidant activity increases have also been linked to improving survival when organisms are challenged against stressful conditions [16–18]. These findings make it clear that HOM treatments can activate a quantifiable biological response on the immune and antioxidant system in juvenile scallop *A. ventricosus*.

	Haemocyte count	Growth rate (μm day^{-1})
Initial	$19 \pm 0.38_e$	
T1	**$44 \pm 0.54_a$**	**$117 \pm 1.32_a$**
T2	$41 \pm 0.57_b$	**$108 \pm 0.97_a$**
T3	$34 \pm 0.75_c$	$53 \pm 6.31_c$
T4	$24 \pm 0.40_d$	$94 \pm 1.38_b$
T5	$31 \pm 0.32_c$	$36 \pm 5.83_d$
T6	$21 \pm 0.87_e$	$14 \pm 0.40_e$
T7	$22 \pm 1.13_{de}$	$34 \pm 0.06_d$
T8	$16 \pm 0.60_f$	$20 \pm 0.14_e$

Numbers show mean ± standard error. Identical lowercase letters denote lack of significant differences between treatments at $p < 0.05$.

Table 1. Haemocyte count and growth rate in *Argopecten ventricosus* spat treated with homoeopathic medicines for 21 days.

Trial 2. To compare between homoeopathy and antibiotic efficiency in *A. ventricosus* juveniles, an experimental design was performed by triplicate (6 treatments and 18 replicates, 120 seeds each replicate) at CIBNOR for 21 days. Juveniles (4.14 ± 0.06 mm; 13.33 ± 0.03) received PaV-Pha (T1), PaV-Sit (T2) as HOM treatments, ampicillin AMP (T3) as antibiotic treatment, and ethanol ET (T4) and NT (T5) as control treatment. Liquid treatments were applied (100 μl l^{-1} for homoeopathy and 10 ppm for antibiotic) after seawater exchange every 48 h and before feeding scallops. Evaluations were performed for growth in height (mm), total wet weight of the shell (mg), biochemical flesh composition (mg g^{-1}) and SOD activity. Thirty juveniles were taken at random from each replica at 7, 14 and 21 days to measure size and total wet weight and to determine absolute growth in height and weight (mm, mg) for each treatment.

Juveniles grew significantly more in size with HOM T1 (6.22 ± 0.11 mm; 0.05 mm d^{-1}) and T2 (6.99 ± 0.09 mm; 0.08 mm d^{-1}) compared with NM (T5) (5 ± 0.02 mm; 0.02 mm d^{-1}). A significant increase in total wet weight was recorded with HOM T2 (41.16 ± 0.35mg; 1.3 mg d^{-1}) compared to NT group T5 (24.33 ± 0.10 mg; 0.5 mg d^{-1}). Survival was 100% in all treatments and their replicates (**Figure 1**).

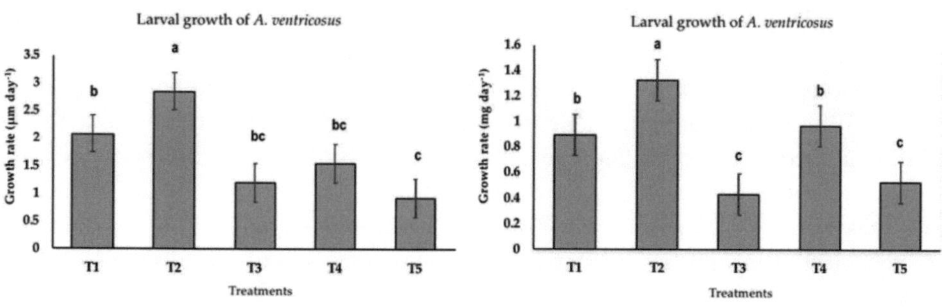

Figure 1. Growth in size (μm day^{-1}) (left) and weight (mg day^{-1}) (right) of juvenile Catarina scallop *Argopecten ventricosus* treated with homoeopathic medicines.

Trial 3. As a continuation of Trial 2, once the previous experiment (21 days) was completed, a pathogen challenge was performed at CIBNOR with those treated juvenile scallops. About 30 juveniles were randomly selected from each previous treatment (in duplicate) and challenged with a pathogenic strain of *Vibrio alginolyticus* (CAIM57: www.ciad.mx). An initial single dose (1×10^7 CFU ml^{-1}) was provided as based on the mean lethal dose (LD_{50}) determined by the Probit method and based on the dose–response model described by Finney [19]. To obtain greater clarity in the results, in addition to the groups previously treated PaV-PhA (T1), PaV-SiT (T2), antibiotic AMP (T3) and ET (T4), two new groups that did not receive any previous treatment were included. First, a new group was infected with CAIM57 and defined as positive control (CTRL +), and another new group was not infected and defined as negative control (CTRL -). Survival (%) of juveniles was evaluated at 0, 24, 48, 72 and 120 h after infection, and the activity of SOD before infection at 48, 72 and 96 h post-infection was determined. For each treatment, soft tissues (100 mg) from six juveniles were weighted and 500 μl phosphate buffer (pH 7.5) were added. The tissues were homogenised and centrifuged at 9327 × *g* for 10 min at 4°C, recovering the supernatant and storing it at −20°C until further analysis. SOD activity was determined with a commercial kit (SOD Assay Kit #19160, Sigma-Aldrich). Results were expressed as an indirect measure of SOD activity as a per cent of the water-soluble tetrazolium salt formazan complex inhibition. During the challenge, no water changes were made. All juvenile scallops not treated but challenged (CTRL +) died at 72 h while untreated and unchallenged scallops attained the highest survival (95%). The HOM-treated scallops also survived the challenge; T1 scallops attained 85 *versus* 40% survival in those treated with antibiotic (T3). Finally, the SOD activity increased significantly with respect to the other treatments and controls in the juveniles of the HOM T1 (81%), 72 h post-infection.

Trial 4. To assess the effects of HOM treatments on the microbial communities of the gastrointestinal tract (GIT) of juvenile *A. ventricosus* an experimental design was applied at CIBNOR during nursery management with five homoeopathic treatments and three controls, four replicates each. Juveniles were placed in recirculating upwelling nursery units (40 l) each one with four PVC upwelling cylinders for 21 days; the following HOM treatments and controls were applied: ViP-ViA/a (T1), ViP-ViA/b (T2), AcF-MsS (T3) PhA-SiT (T4), ViT (T5), ET (T6), and NT (T7). At the end of the experiment, eight scallops were randomly taken from each replica and washed, removed fouling organisms of external sides of shell and sprayed with ethanol and dried. Immediately one of the shells was removed and soft tissues dissected to isolate the gastrointestinal tract (GIT) of each scallop, which were fixed in RNAlater® (Thermo Fisher Scientific, Waltham, MA, USA) and preserved at −20°C. The technique of massive DNA sequencing was applied, which is widely used in the study of microbial communities associated with biological systems. The bacterial 16S rDNA was extracted according to Garcia-Bernal *et al.* [3], amplified for sequencing in the Illumina MiSeq Platform (Illumina, San Diego, CA, USA) in a certified Genomic Services Laboratory (www.langebio.cinvestav.mx; Irapuato, Guanajuato, México). Afterward, a bioinformatic and statistical analysis of the generated database was carried out. Initially, significant differences were detected ($p < 0.05$) in growth rate of shell length (μm d^{-1}) ($p < 0.05$), with the best results (140 μm d^{-1}) in the HOM T2. The groups that received T3 and T5 showed a significantly higher survival rate ($p < 0.05$) than the other groups. Moreover, the dominant phylum was *Proteobacteria*, followed by *Actinobacteria*, *Firmicutes*, and *Bacteroidetes*. The prevalent genera in GIT were *Microbacterium*, *Bacillus*,

Symbiobacterium and *Burkholderia*. In general terms, phyla *Proteobacteria* and *Actinobacteria* play an essential role in immunity and nutrition of invertebrates, and both were dominant in juvenile scallop *A. ventricosus* in all groups treated with HOM.

2.1.2. Horse-mussel (*Modiolus capax*)

The Horse mussel *Modiolus capax* is a native species from the Gulf of California with aquaculture potential, but scientific knowledge must be generated to achieve a sustainable production [20]. A study was developed at CIBNOR to assess the effect of HOM treatments on gonadal, physiological and transcriptomic maturation in *M. capax*. Microalga and wheat meal were used as food for broodstock mussels. The experiment was designed in triplicate (482 adult mussels; 60 days) and applied to evaluate three HOM treatments: SiT-CaS-HeS (T1), PhA-FeP-ZiP (T2), ViP-ViA-ViT (T3), ethanol as positive control (T4), and NT as negative control (T5). A sequential sampling in time (days) was made (t_0, t_{30} and t_{60}). The histological analyses showed that mussels treated with T1 and T3 attained (t_{30}) the best results ($p < 0.05$) in total weight increase, gonadal maturation of 100% females (**Figure 2** left), highest frequency of vitellogenic and postvitellogenic oocytes and highest oocyte quality according to the total area, theoretical diameter and % of ovoplasm. Also, mussels treated with T3 attained the best reproductive condition of the females (gonadal coverage area, ovarian maturity index, gonadal development index and reproductive potential); however, those receiving T2, increased oogonia proliferation and bioenergetic quality of the oocytes (amount of lipids and neutral carbohydrates). Histochemical and biochemical analyses revealed that HOM treatments (T1, T2 and T3) contributed to increase the overall energy reserves (lipids, carbohydrates and proteins) in the ovary, digestive gland and adductor muscle. López-Carvallo *et al.* [20] using wheat enriched di-algal diet, barely reaching gonad to maturity ~ 25% of the *M. capax* broodstock. Thus, we considered that better results in reproductive condition and oocyte quality of the species were attained with HOM treatments with respect to control.

On the other hand, a *de novo* transcriptome characterisation of the ovarian tissue treated with homoeopathy was performed using RNAseq. *In silico* analysis of differential gene expression revealed that mussels treated with T3 showed the highest number of differentially expressed transcripts (**Figure 2** right), and some of them were related to genes that encoded oestrogen

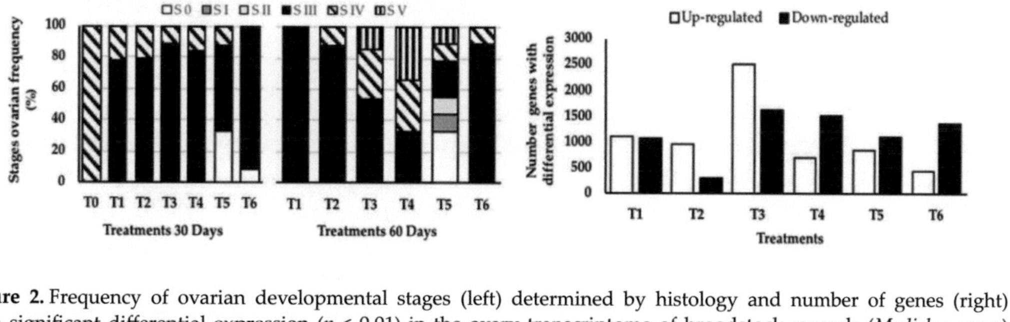

Figure 2. Frequency of ovarian developmental stages (left) determined by histology and number of genes (right) with significant differential expression ($p < 0.01$) in the ovary transcriptome of broodstock mussels (*Modiolus capax*) conditioned with homoeopathy for gonad maturation. Stage S0 undifferentiated; stage SI Previtellogenesis; stage SII Vitellogenesis; stage SIII Posvitellogenic; stage SIV partial spawning; stage SV post spawning; n = 482 mussels. Over-expressed genes = black bars; under-expressed genes = Grey bars.

receptors ER and ERR2- like ($p < 0.01$). Estradiol levels have been reported to have a direct influence on gonadic development and oocyte quality in marine bivalves [21] that allowed explaining the high reproductive condition found in the mussels that received T3.

This work has provided a *de novo* transcriptome characterisation for *M. capax* for the first time and together with the evaluation of physiological variables, it constitutes the first research about the beneficial effect of homoeopathy in gonadal maturation of the species, which has a clear technological applicability during broodstock gonadic conditioning for seed spawning and production of the species. It will likely be applicable to other bivalve molluscs with commercial interest.

2.1.3. American oyster (Crassostrea virginica)

The American oyster *Crassostrea virginica* is an important fishery and aquaculture resource in the Atlantic coast in the Caribbean and in the Gulf of Mexico. A commercial oyster laboratory is already operating in Centro Ostrícola Tecnológico de Tabasco (COTET), Mexico, and homoeopathic medicines were assessed to improve the species culture. HOM treatments were assessed in larval culture and settlement and early nursery of on-cultch and cultch-less seed. Also, a transcriptomic focus study was developed in adult broodstock oyster. The results obtained in six experiments conducted with *C. virginica* are detailed below.

Trial 1. A study was conducted at CIBNOR to analyse the transcriptomic response to five HOM treatments in broodstock oyster *C. virginica* to understand the response mechanisms that are activated by these treatments, thus helping to characterise its mode of action. Adult oysters (120 g; 10–12 cm) from a homogeneous population were conditioned in 80-l plastic boxes provided with continuous microalgal food at COTET for 45 days. An experimental design (3 replicates/treatment; 25 oyster/replicate) was applied with four HOM treatments: ViP-ViA (T1), PhA-SiT (T2), END (T3), INF (T4) and two controls ET (T5) and NT (T6). As treatment vehicle, fully impregnated inert homoeopathic pills were provided one daily per oyster. Routine techniques of the oyster hatchery COTET were applied for conditioning broodstock oyster, and during the trial period, two samplings (S_1 and S_2) were performed for histological analyses at CIBNOR. As a general rule, at the end of the conditioning assay, all conditioned groups were capable of sexually maturing and spawning in greater or lesser percentage, but specific details were observed that allowed to distinguish effects associated to different treatments and controls. Female oysters treated with HOM T1 attained the highest frequency of gonad in vitellogesis (48%) and postvitellogenesis (30.2%) stages at first sampling (T1-S_1) and 16.7% in the second sampling (T1-S_2). In T2-S_2 oysters, the highest frequency (31.6%) showed at resting (undifferentiated) stage and 30.8% post-spawning stage. In ET control T5-S_2, 38.5% of the female oysters were at resting stage and 30.8% at post-spawning stage. Oysters of NT control group T6-S_2 recorded the highest frequency (37.5%) in postvitellogenic stage, 31.3% in partial spawning stage and 25% in post-spawning stage. In the veterinary HOM T3-S_1, oysters showed the highest frequency in partial spawning stage (71.4%), and in T4-S_1, the highest frequency (57.1%) was recorded in post-spawning stage. It is important to highlight that even when T1, T2, T3 and T6 promoted maturation and spawning of the male oysters, there were differences between them related to gamete quality since the presence of atretic oocytes was seen in some samples of T5 and T6 control treatments. The veterinary HOM T4 seemed to have promoted damage to gametes because abundant degenerative oocytes were observed. To evaluate the quality of the oocytes matured under different treatments, the Sudan Black histochemical technique was used staining the lipid components of the cells. The triglyceride lipid index (TLI) was calculated as described by Rodríguez-Jaramillo *et al.* [22], and significant differences were found in oocytes from different treatments and sampling times. Lipid content (TLI) was significantly higher ($p < 0.0001$) in oysters receiving HOM T2 in a short conditioning time (S_1). The rest of the treatments and replicates recorded significantly lower TLI values, including Veterinary HOM T3 and T4. A difference between HOM treatments lies in the fact that some of them seem to trigger a constant production of new generations of oocytes, which may be useful to hatchery purposes because it could be associated to the possibility of several partial spawning events that could derive into several larval batches and more opportunities for seed production.

Trial 2. A study was conducted at CIBNOR to analyse the transcriptomic response to five HOM treatments in broodstock oyster *C. virginica* in order to understand the response mechanisms that are activated by these nanomedicines, thus, helping to characterise their mode-of-action. About 25 groups of 25 adult oysters (120 g; 10 ± 12 cm), each from a homogeneous population, were conditioned in 80-l plastic boxes provided with continuous microalgal food at COTET for 45 days. An experimental design with seven different treatments, each one with three replicates, was developed to determine the transcriptomic effect of an *Actinomycetes* strain (1×10^6 CFU ml^{-1} = RL8) and four homoeopathic drug complexes, alone and in combination: RL8 (T1), ViP-ViA (T2) PhA-SiT (T3), ViP-ViA + RL8 (T4), PhA-SiT+RL8 (T5) and two control groups: ethanol as a positive control (T6) and NT negative control (T7). Five oysters of each homoeopathic and control treatments were initially (t_0) collected and dissected. The rest of the oysters were collected and dissected at the end of the experiment (t_{45}). Several tissues, including mantle, gills, gonad, muscle and digestive gland (DG), were separately placed on snap-frozen tubes in RNA later® (Thermo Fisher Scientific, Waltham, MA, U.S.A.) and stored at −80°C. The rest of tissue portion were fixed in Davidson solution for histological examination. DG tissues fixed in RNA later® (Thermo Fisher Scientific, Waltham, MA, U.S.A.) from five individuals of each experimental group were subjected to RNA extraction, for the transcriptomic analysis, using Illumina Hiseq 2000 platform (Illumina, San Diego, CA, U.S.A.). The comparison of the transcriptome data with the KEGG database indicated that the treatments influenced associated metabolic pathways in gonadal development and maturation including "Developmental process involved in reproduction", "Meiotic cell cycle process", "Steroid metabolic process", "Response to oestrogen "and" Regulation of reproductive process ". These pathways were assigned to the KEGG categories of "Developmental process involved in reproduction" which was among the most important category, indicating the significance of signal transduction systems and endocrine regulation of gonad development and function in *C. virginica.*

The transcriptome of T2 and T3 had 998 and 881 genes with a significantly increased level of expression, respectively, compared to the control ($p < 0.05$). In T4 and T5, there were 748 and 789 genes with an increased level of expression, respectively, compared to a separate control ($p < 0.05$). No genes were significantly differentially expressed under T1 compared to a separate T2, T3, T4, T5 and controls ($p < 0.05$). Genes with increased expression following T2 and T3 were associated with biological processes, including metabolic pathways, ribosomal biogenesis, and transport of nitrogen compounds and anions. Following T4 and T5, genes with increased expression were involved in metabolism processes related with response to stress, maintaining protein expression. Genes with increased expression following T1 were associated with protein kinase A (PKA) signalling that regulates stress responses and mitochondrion degradation. The transcriptomic results obtained allowed us to determine that homoeopathic treatments expressed transcripts associated with complex biological processes, such as reproduction, stress response, cell growth and metabolism of the *C. virginica* oyster.

Trial 3. An experimental design was applied at COTET for oyster larval culture assessment including HOM treatments: SiT (T1), ViT (T2), FeP (T3), MaP (T4) and NT (T5) as a control treatment. The experimental larviculture was performed in 20-l plastic buckets three replicates per treatment with filtered seawater, gentle and continuous aeration and the microalga *I. galbana* and *C. calcitrans* (1:1) as food. Three-day old veliger larvae from a mass production fibreglass tank (25 t) were placed in buckets at initial density of 2 larvae ml^{-1}. Liquid HOM treatments were added daily to culture water (0.01%). The larviculture test finalised when the first pedivéliger larva with a retractable foot was detected but without reaching the final stage of eyed and fully developed pediveliger. Analyses were based on initial and three sequential biometrical data from 30 larvae for each treatment replicate. The HOM larval groups showed better development, survival and setting efficiency in comparison with untreated groups. As seen in **Figure 3**, *C. virginica* larvae that received T1 (162 ± 3.08 μm), T2 (162 ± 2.84 μm) and T4 (162 ± 3.13 μm) reached the largest sizes but T5 (145 ± 2.97 μm) reached the lowest sizes, while those larvae receiving T2 reached the highest setting efficiency (83 ± 7.98%) and the highest seed survival (36.5 ± 4.9%).

Trial 4. To assess larval settlement, mature eyed pediveliger larvae were obtained by selective screening of the same bulk culture tank; then, they were placed in 20-l buckets at a seeding density of 0.25 larvae ml^{-1}. The same HOM treatments: SiT (T1), ViT (T2), FeP (T3), MaP (T4) and NT (T5) were applied at COTET during the on-cultch setting process. Seawater change, aeration and microalgae were provided as food, and 50 clean oyster shells/buckets were placed as a natural settlement (cultch) substrate. Liquid homoeopathic treatments were added daily to the culture water (0.02% v/v). The results achieved are shown in **Figure 3**.

Trial 5. A pilot experimental design in triplicate was applied at COTET for larval culture: MaP–FeP (T1), SiT–ViT (T2) and NT (T3) as control. Four-day old larvae were placed in nine-conical fibreglass tanks (750 l) at initial density of 5.33 larvae ml^{-1} and the microalgae *I. galbana and C. calcitrans* (1:1) was provided as food. Larval growth and harvest of pediveliger larval

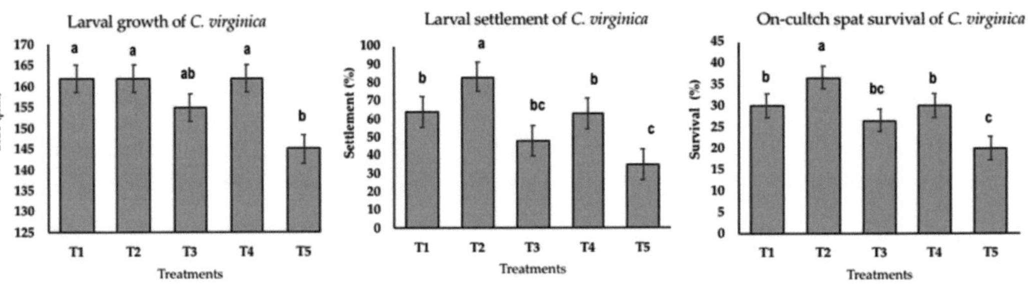

Figure 3. Larval growth (left), larval setting efficiency (Centre) and seed survival in 20-l plastic buckets (right) of American oyster *Crassostrea virginica*, treated with homoeopathic drugs in COTET oyster hatchery.

biomass was evaluated. Temperature and salinity were maintained in the range of 27 ± 3°C and 22 ± 7 psu. Liquid HOM treatments were added daily to culture water (0.01%). Based on biometrical data of 30 larvae for each replicate every 48 h, the best results in larval growth were obtained with HOM T1, followed by control (T3) and finally by T2 (**Figure 4**). The highest yield in biomass of mature pedivéliger larva (9.29 g) was obtained with HOM T2, followed by control T3 (7.42 g) and the lowest production (7.04 g) was obtained with T1 (**Figure 4**). Differential results obtained in this study were attributable to better survival of larvae because homoeopathy favours nutrition and assimilation of nutrients, increases stress resistance because of high culture density and strengthens the immune system and resistance to attack by pathogens [5, 23]. A better survival associated to small sized larvae has been observed in other species treated with homoeopathy, such as *Panopea globosa* and *Crassostrea sikamea* (Mazón-Suástegui *et al.*, unpublished).

Trial 6. An experimental design was applied at COTET for cultch-less spat culture with two homoeopathic treatments and one control (four replicates each) in recirculating upwelling nursery units (40 l) each one allocating four upwellers of PVC pipe 4″ and airlift devices. As HOM treatments, MaP–FeP (T1) SiT–ViT (T2) and NT control (T3) were assessed. Initial seed density was set on 4750 spat/upweller and *I. galbana* and *C. calcitrans* (1:1) was used as food. The total harvest of seed per treatments and replicates (volume) were measured weekly with a graduated cylinder. Temperature and salinity were maintained in the range of 27 ± 3°C and 24 ± 3 psu. Liquid dynamisation treatments were added daily to culture water (0.02%). Based on initial and weekly biometrical data of 30 spat for each replicate, the best growth results were obtained in the control T3, followed by HOM T2 and finally T1 (**Figure 5**). In contrast, the best results in oyster seed biomass were attained with HOM T1 (268 ml) and T2 (242 ml) and the lowest production (236 ml) with control T3 (**Figure 5**).

Oyster seed receiving HOM treatments grew less in regard to individual size, but an overall greater volume of marketable juveniles was harvested, which means that survival was greater in HOM treatments. From a commercial hatchery perspective, it is more important to produce live seed biomass even if small sized than dead ones. In real production and profitability terms, the hatchery could produce and sell more seeds if homoeopathic drugs, such as those evaluated in this study were routinely applied. The results in larval settlement and juvenile

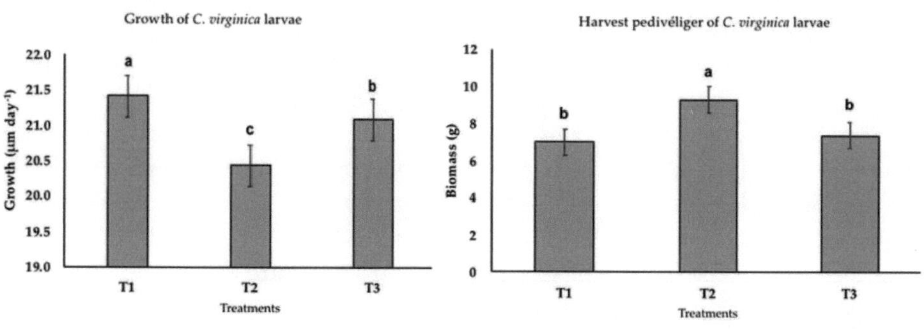

Figure 4. Growth in size (µm) of veliger larvae (left) and harvest of pediveliger larvae (right) of American oyster *Crassostrea virginica* treated with HOM in 750-l conical fibreglass tanks in COTET hatchery.

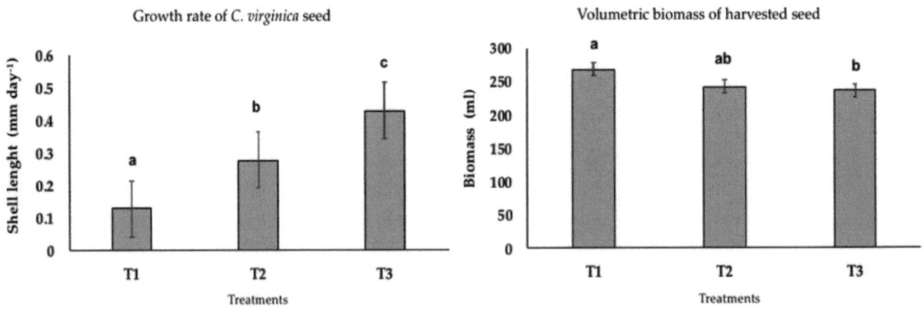

Figure 5. Growth rate in shell length (μm day^{-1}) (left) and volumetric biomass of harvested seed (right) of American oyster *Crassostrea virginica* seed, receiving HOM treatments in recirculating upwelling nursery units (40 l) at COTET hatchery.

nursery confirmed a great potential for aquacultural homoeopathy to strengthen the state-of-the-art technology in oyster seed production and increase actual productivity and economic profitability of the commercial production of the species in the hatchery.

2.1.4. Kumamoto oyster (Crassostrea sikamea)

Kumamoto oyster *Crassostrea. sikamea* is a highly valuable mollusc species and cultivated in several countries, so it is important to generate new knowledge for its culture and alternative treatments for its management in the laboratory. Four experiments with *C. sikamea* were performed as described below.

Trial 1. An experimental design was applied at CIBNOR for conditioning broodstock with two replicates (25 oysters each) in eight plastic boxes (60 l), in which the following treatments were tested: ViP-ViA (T1) and PhA-SiT (T2) as HOM treatments, and ET (T3) and NT (T4). As treatment vehicle, fully impregnated inert homoeopathic pills were used, one-a-day per oyster.

Oysters were dissected (30 at t_0 and 15 from each replicate at t_{42}) and soft tissues histologically processed by haematoxylin and eosin staining to determine gonadic (GI) and digestive gland (DG) indexes, reproductive stages and theoretical diameter in oocytes (DT) according to Rodriguez-Jaramillo *et al.* [22] and Barber and Blake [24]. Moreover, significant differences were observed ($p < 0.05$) in GI in broodstock oysters with respect to the initial value (39%), and the best results were observed in oysters treated with T2 and T3 (63 and 67%). The highest DI was recorded at the beginning of the experiment (60%) and the lowest ($p < 0.05$) for T2 and T3 (36 and 32%), which suggested a greater energy storage in the gonad for oocyte maturation [24] (**Figure 6**). A high frequency of organisms in post-spawning and gonadic development stage (38 and 52%) was observed at t_0. At t_{42}, the oysters of T2 and T3 showed the highest percentage of full sexual maturity (50%) while the greatest number (35 and 40%) of organisms in gonadic development stage was observed in T3 and T4 (**Figure 6**).

Relative to the final DT (35 μm), which was 28 μm at t_0, no significant differences were recorded between HOM-treated and control oysters. However, maturity and undifferentiated stages were simultaneously observed in T3 oysters. Since *C. sikamea* matures and partially spawns several times during the same reproductive season, this result suggested a positive effect by

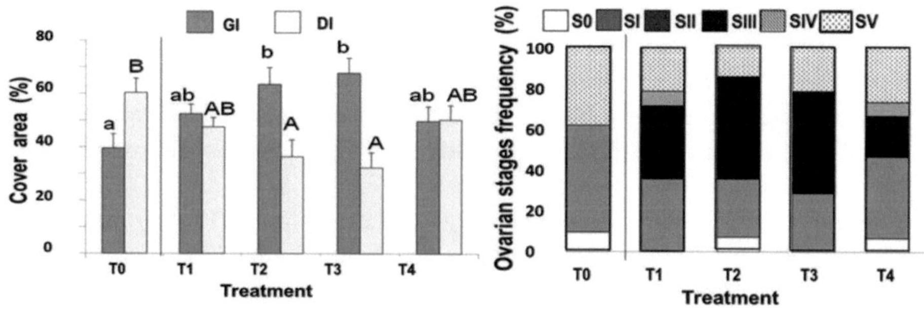

Figure 6. Gonadic index (GI) and digestive index (DI) (left) and gonad development (tight) in broodstock oyster *Crassostrea sikamea* conditioned in the laboratory for 42 days. Different letters show significant differences between treatments ($p < 0.05$). Stage 0: Undifferentiated; stage I: Previtellogenesis; stage II: Vitellogenesis; stage III: Posvitellogenic; stage IV: Partial spawning; stage V: Post spawning.

the activation of a new oocyte production cycle associated to HOM T2 (PhA-SiT). In bivalve molluscs, an expenditure of digestive gland reserves is associated to gonad development and maturation of gametes [24], and T2 was associated to a greater number of sexually mature oysters with higher GI and lower ID values. It could be applicable to gonadic conditioning for sexual maturation of broodstock oyster at the hatchery.

Trial 2. Once the broodstock assay finished, a subsequent second bioassay was developed to evaluate larval performance in two different larval cohorts from HOM-treated spawners (T1, T2, T3, T4; three replicates each) and NT spawners (T5, T6, T7, T8; three replicates each). The greatest growth in length was recorded in T5 and T6 (217 and 212 μm) and the lowest in T3 and T4 (197 and 190 μm). A similar growth pattern was observed in height: the greatest in T5–T8 and the lowest in T1–T4. In general, larvae from untreated broodstock grew more. Nevertheless, as a counterpart, the general mean survival of larvae from HOM-treated spawners (T1-T4) was 32.3% higher than those from NT broodstock, and it could be very important to improve hatchery seed production. Those differences in larval survival suggested a favourable effect of HOM treatments in oocyte quality and then a potential effect in the progeny.

Trial 3. A 35-day experimental design in triplicate (six treatments and 18 replicates, 120 seed each replicate) was applied at CIBNOR to compare efficiency of antibiotics against homoeopathy in juvenile *C. sikamea*. Juveniles (6.13 ± 0.16 mm) were placed in 2-l plastic containers with filtered (1 μm) and UV-sterilised seawater (23 ± 1°C; 38 psu), continuous aeration and microalgal food. Juvenile oysters received PaV-Pha (T1), PaV-Sit (T2), Pav-MeS (T3) as HOM treatments, ampicillin AMP (T4) as antibiotic treatment, and ethanol ET (T5) and NT (T6) as control treatments. Liquid treatments were applied 100 $\mu l\ l^{-1}$ for homoeopathy and 10 ppm for antibiotic after exchange of seawater every 48 h and before feeding oysters. Evaluations were performed on growth in height (mm), and total wet weight of the shell (mg), biochemical flesh composition ($mg\ g^{-1}$) and SOD activity. The biochemical composition was determined by triplicate sampling at the beginning and end of the trial. Samples were stored at −80°C and then lyophilized, rehydrated in 3 ml cold saline solution (35%) and homogenised to obtain crude extracts. Crude extracts were processed at CIBNOR laboratory applying traditional and certified techniques, which are described by López-Carvallo *et al.* [20] and Mazón-Suástegui *et al.* [23]. For SOD analyses, after deep cleaning of shell, soft tissues were dissected and fixed individually in RNAlater®

(Thermo Fisher Scientific, Waltham, MA, USA) at a 1:5 ratio (100 mg tissue: 500 μl RNA-Later®) and then preserve it at −20°C. SOD analyses activity was determined with a commercial kit (SOD Assay Kit #19160, Sigma-Aldrich). Results were expressed as an indirect measure of SOD activity as a per cent of the water soluble tetrazolium salt formazan complex inhibition. The juveniles grew significantly more with HOM T1 (9.27 ± 0.18 mm, 0.073 mm day^{-1}) and T3 (9.36 ± 0.18 mm, 0.076 mm day^{-1}) compared with NT (T6) (8.02 ± 0.23 mm; 0.053 mm day^{-1}), but of all treatments, the ET T5 group was the best (10.17 ± 0.31 mm; 0.105 mm day^{-1}). A significant increase in total wet weight was recorded with HOM T2 (107.33 ± 6.9 mg, 2.0 mg day^{-1}) compared to NT group T6 (76.11 ± 2.8 mg, 1.1 mg day^{-1}) (**Figure 7**). Survival was 100% in all treatments and their replicates. Moreover, the biochemical composition of the juveniles showed significant differences in carbohydrates and lipids but not in proteins. The highest amount of lipids was obtained with HOM T2 (96.32 ± 1.18 mg g^{-1}) and that of carbohydrates in antibiotic T4 (27.48 ± 2.27 mg g^{-1}) compared with the NT (T6) (6.96 ± 1.47 mg g^{-1}). At the end of this trial, SOD activity was higher in HOM T1 (92%) than control T6 (88%) and antibiotic T4 (84%).

Trial 4. Once the previous experiment was completed, a pathogen challenge was made with treated seeds. About 30 juveniles were randomly selected from each previous treatment (in duplicate) and challenged at CIBNOR with a pathogenic strain of *V. alginolyticus* (CAIM57: www.ciad.mx). An initial single dose (1 × 10^6 CFU ml^{-1}) was given as based on the mean lethal dose (LD_{50}) determined by the Probit method based on the dose–response model described by Finney [19].

To obtain greater clarity in the results, in addition to the groups treated with homoeopathy, antibiotic and ethanol, two new groups were included, which did not receive any previous treatment; one of which was infected with the pathogen (CTRL +) and another one that was not infected with the pathogen (CTRL -). Survival (%) of juveniles was evaluated at 0, 24, 48, 72 and 120 h after infection and the SOD activity (using a commercial kit; SOD Assay Kit #19160, Sigma-Aldrich) of before infection and at 2, 24, 48 and 72 post-infection was determined. After being infected with the pathogen, all the juveniles survived and no significant differences were observed between treatments and controls with respect to SOD activity. Up to date, that result has no coherent explanation yet; unless the species is highly resistant, the pathogenic strain has not got sufficient virulence or a wrong (lower) dose was applied because 1 × 10^7 CFU ml^{-1} was applied to juvenile scallop *A. ventricosus* by Mazón-Suástegui [23].

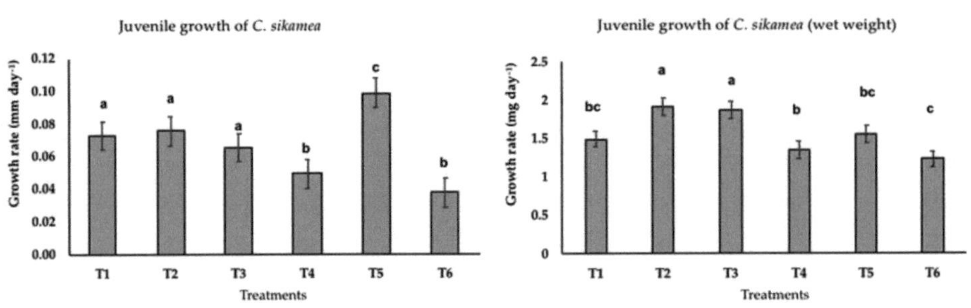

Figure 7. Growth in size (mm day^{-1}) (left) and total wet weight (mg day^{-1}) (right) in juveniles of the oyster Kumamoto *Crasssostrea sikamea* treated with homoeopathic medicines.

2.1.5. Geoduck clam (*Panopea globosa*)

The geoduck clam *Panopea globosa* is an important marine resource distributed on both coasts of the peninsula of Baja California, Mexico. Its cultivation is still in experimental stage and totally depending on juveniles produced in the laboratory. As with other bivalves, there are limitations in their production due to mortalities of larvae and seeds, associated with the presence of pathogens. This problem has led to the search for new eco-friendly alternatives such as aquaculture homoeopathy, which has a positive effect on nutrition, health and immune response of bivalve molluscs, shrimp and marine fish [5, 23]. Our study evaluated the effect of various HOM treatments on growth, survival and microbiota of the gastrointestinal tract (GIT) of the species. Juvenile (spat) *P. globosa* with an average length of 1.98 ± 0.1 cm were produced in the laboratory and provided by the company Acuacultura Robles, a commercial mollusc hatchery located in La Paz, B.C.S. Mexico. Clams with an average length of 1.98 ± 0.1 cm were produced and provided by the company Acuacultura Robles, acclimatised at CIBNOR and then cultured (21 days) in nursery units previously described for *A. ventricosus*. About 24 upwelling units were used, each one with 52 clams and 13 clams per upweller cylinder; the following HOM treatments and controls were applied: ViP-ViA/a (T1), ViP-ViA/b (T2), AcF-MsS (T3) PhA-SiT (T4), ViT (T5), ET (T6) and NT (T7). Samples were taken at the beginning (t_0) and end of the experiment (t_1) by randomly selecting eight clams per replica, accounting for a total number of 216. After external deep cleaning, soft tissues were dissected to isolate GIT of each clam to fix individually in RNAlater® (Thermo Fisher Scientific, Waltham, MA, USA) at a 1:5 ratio (100 mg tissue: 500 μl RNA-Later®) and then preserve it at −20°C. The bacterial 16S rDNA was extracted according to Garcia-Bernal *et al.* [3] and amplified using Illumina MiSeq Platform (Illumina, San Diego, CA) in a certified Genomic Services Laboratory (www.langebio.cinvestav.mx; Irapuato, Guanajuato, México). Afterward, a bioinformatic and statistical analysis of the generated database was carried out. Taking as reference the microbial diversity in the GIT of *P. globosa* juveniles, the best results were obtained with HOM T1, a nosode product developed at CIBNOR. T1 favoured dominant abundance of the Proteobacteria phylum and some of its classes as γ-Proteobacteria. In that sense, similarities were observed with the microbiota of other marine species, and that part of the microbiota found in *P. globosa* is associated with stimulation of the immune system. Overall, the results indicated that the HOM treatments modified the abundance of the microbial communities of the species, mainly in the phylotypes related to nutritional processes. On the other hand, significant differences were recorded with respect to growth in weight and length ($p < 0.0001$) between the clams that received HOM treatment and the control groups. The highest growth in weight was recorded in T3, T4 and T5. The difference in growth in length was smaller but equally superior to these HOM treatments. Significant differences ($p = 0.019$) in survival were also observed. The highest value (95%) was recorded in clams with HOM T3 followed by the NT control group T7 (93%) without homoeopathy or ethanol. In contrast, the lowest survival (76%) was observed in T2 and also in the ET control group T6.

2.1.6. Octopus (*Octopus bimaculoides*)

An experimental design in triplicate (10 juveniles/replicate) was developed in 60-l fibreglass units at CIBNOR laboratory to assess growth and survival in juvenile octopus during a 28-day period. This assay was intended to study and compare the effects of fresh crab *Callinectes belicosus* and squid *Dosidicus gigas* meat, as raw or thermically processed food (35°C and 60°C). Also, a HOM treatment was added to culture water, as a digestive system enhancer to *Octopus bimaculoides*. Two processed food treatments (35°C and 60°C), two HOM treatments (HOM-35 and HOM-60), and a positive control treatment (unprocessed raw food) were assessed. As HOM treatment, PhA-SiT was added directly to culture water alternating each medicine every day from Monday to Saturday. Food was provided *ad libitum* once a day. Wet weight (day 0 and 17) and survival percentage (day 17) were recorded for all 15 groups (**Table 2**). As expected for a positive control with a traditionally used raw food, the highest survival was attained in juvenile octopus fed on raw meat (97%) but also with HOM-35 (93%) and HOM-60 (86%) the lowest survival was seen in octopuses fed processed food 60°C without giving them HOM treatments (**Table 2**).

Knowledge concerning octopus culture is recent and scarce even when completing the life cycle successfully in captivity which is possible with some species [25]. One of the main obstacles to achieve production level is the lack of an industrialised food to be physiologically and economically viable since to date the only efficient food is fresh flesh or live preys [26]. In meal production, raw material goes through aggressive thermal processes that induce protein denaturalisation, carbonylation, hydrophobicity and aggregation [27]. Octopus digestive enzymes are sensitive to these effects; therefore, they cannot hydrolyse their substrates, reducing food digestibility and octopus growth [28, 29].

Heat treatment to raw crab and squid meat to obtain meal ingredients to formulate a balanced diet is not traditionally preferred because this process denaturalises proteins, reduces digestibility and assimilation [28], and lipids can oxidise [30]. Experimental results suggested an enhancement of enzymatic function in *O. bimaculoides* promoted by HOM (PhA-SiT) treatment dissolved into culture water when food was not processed. HOM treatment seemed to have increased the digestive capability in juvenile octopus and the assimilability of processed food (35°C), but the loss of quality in crab and squid meals in 60°C food could not be compensated by the HOM treatment.

2.2. Crustaceans

2.2.1. White shrimp (Litopenaus vannamei)

The White shrimp *L. vannamei* is a species with rapid growth, high survival and price in the market, which makes this crustacean one of the most important resources at worldwide level.

Item.	HOM-60°C	60°C	HOM-35°C	35 °C	No HOM
W Day 1 (g)	0.11 ± 0.05	0.10 ± 0.03	0.10 ± 0.04	0.12 ± 0.04	0.12 ± 0.04
W Day 28 (g)	0.11 ± 0.06	0.11 ± 0.06	0.15 ± 0.06	0.20 ± 0.06	0.20 ± 0.07
Survival (%)	28	46	87	54	93

Data are presented as Mean value ± SD. W = Wet weight.

Table 2. *Octopus bimaculoides* juvenile growth and survival during food and HOM treatment assessment.

Nonetheless, the production of this important resource has been hindered by recurrent epizootic outbreaks and sudden mortalities caused by pathogen microorganisms. To face the challenge, several chemical and antibiotic products have usually been applied whose prophylactic application was initially an effective strategy. However, they have caused the development of resistant bacteria making it necessary to reduce their application. These problems have led the shrimp industry to explore and develop new and more strategies, as effective as or better than antibiotics, eco-friendly and with long-term sustainability. Previous studies have indicated that homoeopathic medicines stimulated the immune system and caused specific organic responses [31, 32]. Therefore, we evaluated the effect of homoeopathic medicines in growth and survival of *L vannamei* postlarvae under controlled laboratory conditions (Trial 1); the survival and antioxidant response through superoxide dismutase (SOD) activity in juveniles when challenged (1×10^6 CFU ml^{-1}) with a pathogenic strain of *V. parahaemolyticus* (CAIM-170) at CIBNOR, in a biosecurity laboratory (Trial 2); during the production of postlarvae in a commercial hatchery, focusing on dynamics of the bacterial populations (Trial 3); zootechnical (growth) results and gene expression (Trial 4); and growing-out to marketable size in a commercial farm (Trial 5).

Trial 1. To evaluate growth and survival of postlarvae, an experimental design was applied at CIBNOR with three HOM treatments: ViP-ViA (T1), PhA-SiT (T2), ViP-ViA + PhA-SiT (T3) and ET (T4) as control. HOM treatments were applied for 30 days, spraying liquid dynamisations in commercial pelleted food, administered *ad-libitum*. In general, the best results were obtained in T3 (T1 + T2), showing a clear synergy between T1 and T2 (**Table 3**).

Trial 2. To assess survival and SOD activity in juveniles, an experimental design with four HOM treatments: INM (T1), PaV (T2), INM-PaV (T3), ViT (T4) and NT (T5) was applied at CIBNOR. HOM treatments were applied to juveniles 7 days prior to challenge and 5 days during challenge. Liquid dynamisations were sprayed in commercial pelleted food, which was supplied *ad-libitum*, 7 days prior to and during challenge. At 70 h after the start of the challenge, SOD was determined in shrimp tissue. At the end of the challenge (120 h post-infection), the shrimp treated with T2, T3 and T4, exhibited significantly higher average survival ($p < 0.05$) than the control group T5. Juveniles treated with T3 and T4 showed the highest

Parameter	Treatments			
	T1	T2	T3	T4
LI (cm)	**4.64 ± 0.94 b**	4.61 ± 0.94 a	3.92 ± 0.93 a	4.59 ± 0.70 a
WI (g)	2.82 ± 0.60 a	**2.92 ± 0.65 b**	**3.15 ± 0.74 b**	2.62 ± 0.53 a
DWI (g day^{-1})	0.018 ± 0.007 ab	0.02 ± 0.009 ab	**0.052 ± 0.018 c**	0.016 ± 0.007 a
GR	0.094 ± 0.020 a	0.097 ± 0.021 b	**0.105 ± 0.024 c**	0.087 ± 0.017 a
S (%)	99.30	97.33	**99.33**	98.00

Different letters in the same column, indicate significant difference ($p < 0.05$). Length increase (LI), weight increase (WI), daily weight increase (DWI), growth rate (GR), survival (S).

Table 3. Growth and survival of *Litopenaeus vannamei* postlarvae treated with homoeopathic treatments during an experimental assay at CIBNOR laboratory.

survival with 64.43 and 56%, respectively, while all those treated with T1 and T5 died. These results suggested that the greatest survival of HOM-treated shrimp could have been related to a stimulation of their immune system, and consequently, to a greater resistance to acute infectious diseases associated with the genus *Vibrio* [32]. Regarding SOD activity, at 70 h after the start of the challenge, the groups treated with T1 and T2 did not show significant differences (34.48 ± 1.87 and 16.32 ± 1.22, respectively) in relation to the control group (41.63 ± 2.59) while T3 and T4 with a SOD activity of (86.43 ± 1.02 and 83.47 ± 5.54, respectively) exhibited significantly higher values than the control group ($p < 0.05$) (41.63 ± 2.59) (**Figure 8**).

Treatment T3 contained PaV used in human medicine as a tranquilliser to reduce stress and improve sleep, in addition to Heel-Mix (Rubiopharma®, Mexico) that contains medicines used for the treatment of enzymatic disorders, infectious diseases and stimulation of the body defences in human beings, whereas T4 is a homoeopathic medicine whose active principle is the venom of the scorpion *Rhopalurus junceus* which is an endemic species of Cuba. These results suggested a potential applicability of the homoeopathic medicines studied as prophylactic treatments to reduce stress and to improve shrimp immune response, which could help reduce the incidence of epizootic diseases and massive mortalities that have been a great problem for the industry due to economic losses associated with *V. parahaemolithycus* in shrimp culture.

Trial 3. A third experiment consisted of the evaluation of HOM treatments during the commercial production of *L. vannamei* postlarvae in the facilities of the commercial hatchery Aquacultura Mahr in square concrete tanks with a capacity of 20 t. An experimental design was applied with four tanks with three HOM treatments: MaP-CaP-Fep-Zip-PhA (T1) Hes-Sit-Cap-Pha (T1), INM-PaV-ViT-PhA (T3), and NT as the control group (T4). Liquid homoeopathic dynamisations were added to commercially pulverised food given periodically for a 24-h period (day-night). Shrimp farming has been affected by viral and bacterial diseases principally those associated to a highly virulent strain of *V. parahaemolyticus*; this problem is of primary interest concern [33], so this assay was intended to determine the population dynamics of the bacterial populations mainly *Vibrio* spp. Bacteriological examination of isolated vibrio species depend mainly on using TCBS agar as a selective media to differentiate

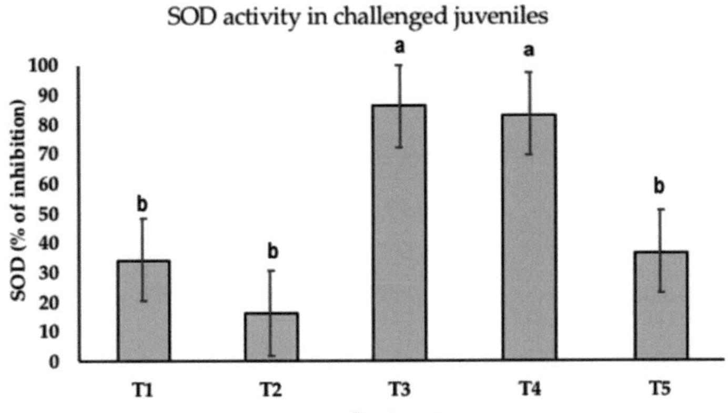

Figure 8. SOD activity in HOM-treated juvenile shrimp *Litopenaeus vannamei* treated with homoeopathy and then challenged with *Vibrio parahaemolithycus*.

between sucrose and non-sucrose fermented colonies; *V. alginolyticus* is sucrose fermenter and shows yellow-coloured colonies while *V. parahaemolyticus* and *V. vulnificus* are non-sucrose fermenters and have green colonies [34]. Although no statistical significant ($p > 0.05$) differences were found in HOM-treated and non-treated PLs, positive effects were apparently attained with HOM treatments (**Figure 9**). Research must continue not only with controlled and multi-replicate experimental designs in the laboratory but also efforts should continue in commercial hatcheries even if obvious difficulties exist because production is at most the first priority and not necessarily compatible with a strict and traditional scientific research.

Trial 4. A fourth assessment was made also at Aquacultura Mahr hatchery. An experimental design was applied in six concrete tanks (20 t) with five HOM treatments: BaC-INM, Sit-INM (T2), HeS-INM (T3), PhA-INM (T4), PaV-INM (T5), and two NT control groups (T6 and T7). Liquid homoeopathic dynamisations were provided with pulverised food as vehicle. Growth in weight of the shrimp postlarvae was exponential with a correlation coefficient (r^2) > 0.95. The best treatment was T5 (PaV-INM). The NT control group had the lowest growth rate and the lowest survival (20.2%), while in the HOM-treated groups, it was 25.6 ± 5.38% (21–34.1%) (**Figure 10**, Left). At the end of the production cycle, gene expression analyses were made at CIBNOR to compare HOM-treated with non-treated postlarvae. The results of gene expression related to the activities of the aminopeptidase (AMP), amylase (AMY), chymotrypsin (CHY) and trypsin (TRY) enzymes showed clear and statistically significant differences ($p < 0.05$; $n = 30$) between the HOM-treated and NT postlarvae (**Figure 10**, Right). These results confirmed a positive impact of the use of homoeopathic medicines in the commercial production of *L. vannamei* postlarvae.

Trial 5. A fourth experiment consisted of the evaluation of HOM treatments during the mass cultivation of *L. vannamei* from postlarvae to adult size in commercial facilities of the company BCS Camarón, a commercial shrimp farm. The evaluation was performed in six earthen ponds of 10 ha each, initially seeded at a density of 8 PL m^{-2}. An experimental design was applied with six earthen ponds 10 ha, four ponds with HOM treatment and two NT ponds without homoeopathy: PhA-SiT (T1), PaV-ViT (T2), and NT as negative control group (T3). Liquid homoeopathic dynamisations were sprayed in commercially balanced food and applied in the culture ponds, each treatment component on alternate days for 130 days. Three samplings were made: after seeding and 7 days for acclimatisation (t_1); after 52 days (t_{52}) and after 130 days post seeding (t_{130}). Growth parameters and biomass production as body weight (BW), total length (TL) and weight gain average (WGA) were determined. As physiological health indicators, the hepatopancreatic coverage index (HCI) was

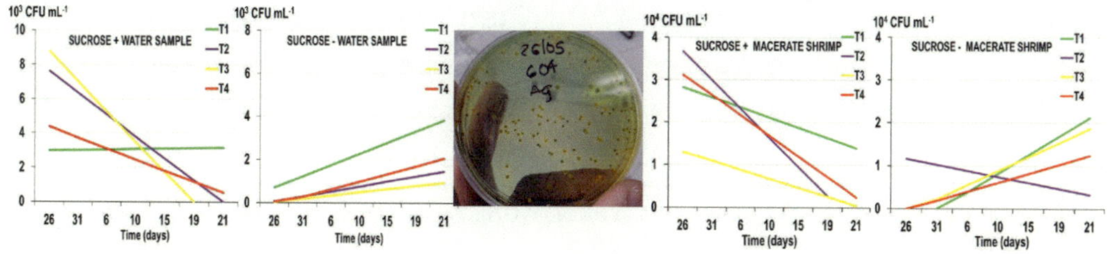

Figure 9. Straight regression showing the increasing trends in the bacterial count of *Vibrio spp*. in culture seawater and macerate of *Litopenaeus vannamei* postlarvae. Petri dish with TCBS culture medium, which was previously inoculated, showing the presence of sucrose positive (yellow) and sucrose negative (green) colonies of the genus *Vibrio*.

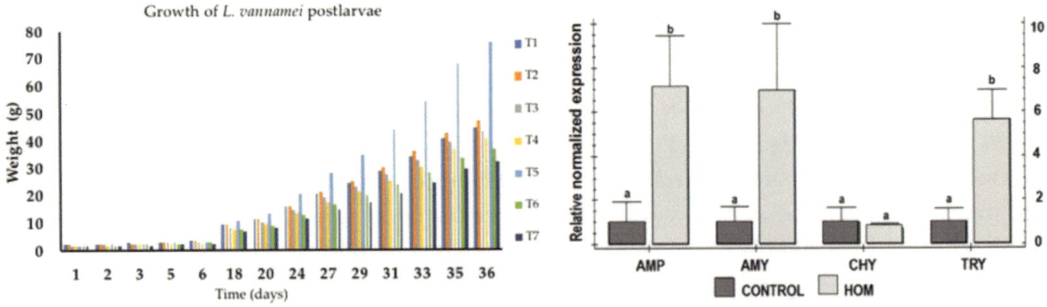

Figure 10. Growth in weight (left) and gene expression (right) in *Litopenaeus vannamei* postlarvae or adult treated and with homoeopathic medicines in a commercial shrimp hatchery (Acuacultura Mahr). Aminopeptidase (AMP), amylase (AMY), chymotrypsin (CHY), trypsin (TRY).

evaluated as a morpho-histological index variable [HCI (%) = hepatopancreatic coverage area/cephalothorax coverage area × 100]. From an initial time characterised by no significant differences in BW and TL after 130 days of culture both HOM (T1 and T2) produced the best growth results in BW, TL and WGA compared to NT ponds (**Table 4**).

The shrimp treated with PaV-VIT after 52 days of treatment (T1) achieved the highest HCI (32.32 ± 0.61%) and differed from PhA – SIT and un-treated control ponds. (28.34 ± 0.87%; 26.58 ± 0.64%, respectively). At the end of the experiment (130 days), both homoeopathic treatments showed better HCI; in relation to the negative control group (PhA – SIT = 33.87 ± 1.02%; PaV – VIT = 33.31 ± 0.77% control T3 = 26.54 ± 0.56%) as shown in **Figure 11**.

The positive effect of the HOM treatments evaluated in growth and morpho-histological index of *L. vannamei* could be attributable to smaller micelle and higher activity in water with

Growth variables	Experimental groups (Two 10 h ponds each)		
	T1	T3	T2
BW T_0 (7 days)	2.47 ± 0.06 ^{a}	2.43 ± 0.06 ^{a}	2.52 ± 0.07 ^{a}
BW T_1 (52 days)	16.79 ± 0.22 ^{a}	16.02 ± 0.19 ^{a}	**18.83 ± 0.21** ^{b}
BW T_2 (130 days)	20.08 ± 0.18 ^{a}	18.59 ± 0.17 ^{b}	**23.90 ± 0.29** ^{c}
TL T_0 (7 days)	7.13 ± 0.05 ^{a}	7.08 ± 0.05 ^{a}	7.06 ± 0.06 ^{a}
TL T_1 (52 days)	13.89 ± 0.05 ^{a}	13.71 ± 0.06 ^{a}	14.31 ± 0.06 ^{b}
TL T_2 (130 days)	15.02 ± 0.05 ^{a}	14.66 ± 0.05 ^{b}	**15.81 ± 0.06** ^{c}
WGA T_1 (52 days)	14.37 ± 0.27 ^{ab}	13,59 ± 0.05 ^{a}	15,72 ± 0.78 ^{b}
WGA T_2 (130 days)	17.58 ± 0.26 ^{a}	16,41 ± 0.14 ^{b}	**21,08 ± 0.64** ^{c}

Body weight (BW), Total length (TL), weight gain average (WGA). Values within the same row with different letters represent significant differences ($p < 0.05$).

Table 4. Growth of *Litopenaeus vanamei* treated with homoeopathy while cultured in semi-intensive system in a commercial shrimp farm (BCS Camarón Farm; six earthen ponds, 10 ha).

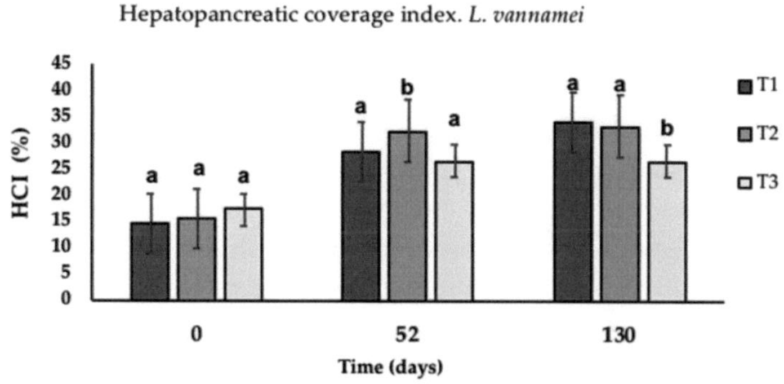

Figure 11. Hepatopancreatic coverage index (HCI %) of adult shrimp in a commercial shrimp farm (earth ponds 10 h) treated with homoeopathic medicines.

nanoparticle content [35]. The presence of nanoparticles has been demonstrated in highly homoeopathic ultra-diluted medicines [8]. On the other hand, the same homoeopathic medicines used in shrimp ponds, (PaV, PhA, SIT and ViT) outperformed antibiotics in juvenile scallop *A. ventricosus* [23] These results enhanced the evidences that aquacultural homoeopathy has applicability in shrimp commercial culture to improve the productivity of the shrimp industry.

3. Conclusion

One of the most promising and novel fields that strengthen the immune system in marine organisms is the use of immunostimulants that are natural compounds modulating the immune system and increasing resistance of the host against disease mainly those caused by bacteria [18]. Regarding immunostimulants, homoeopathy has been proposed as a novel alternative in aquaculture practices to improve health and strengthen the organism's immune response [5, 23]. The production of marine mollusc, shrimp and fish around the world still challenged every year by the propagation and emergence of new diseases, mainly those related to viruses and bacteria such as *Vibrio spp.* and *Aeromonas spp.*, which are treated with conventional methods as antibiotics [2, 14]. The findings in our research suggested that homoeopathic medicines have a great potential to increase health and performance in marine mollusc including bivalves and octopus, shrimp and marine fish.

Findings in the scallop *A. ventricosus* support the fact that homoeopathic medicines do not act directly over the disease-cause *per-se*, killing the bacteria or removing the stressing agent but enhancing the capacity of the HOM-treated organisms to resist the infection or overpass stressful conditions. Some results in molluscs have shown a greater effect of homoeopathic medicines on survival than on growth of larvae and seeds. However, it is necessary to consider that greater survival implies maintaining a higher density in larval culture and that not only larvae of larger size are able to settle and become marketable seeds with a good performance in the field. From a commercial point of view, it is more important to attain bigger biomass of successfully setting larvae even if small sized because independently of their size, all larvae are capable and competent for setting process and seed sales. This work has also contributed to the knowledge of octopus for aquaculture purposes since this organism is an important fishery in Mexico with precocious development and high growth rates. For this and all other species, homoeopathic medicines are not intended as food additives; on the contrary, they contribute to acquire a better internal homeostasis, and as a consequence, a better digestive enzyme function and nutrition, and an enhanced immune system, despite of lacking differences in growth parameters. Regarding fish culture, this industry is increasing with time in Mexico but especially marine fish, which represents a great opportunity to improve fish culture by using homoeopathy; as it was demonstrated in this work, it participates at diverse developmental stages, enhancing fish health and growth performance.

To date, overall results are positive and suggest that homoeopathy is a natural, viable and eco-friendly treatment to reduce the use of disinfectants and chemotherapeutics, including antibiotics, in mollusc, shrimp and marine fish industries, to reduce stress, improve nutrition and immune response, to increase their resistance to any of the various pathogenic strains of bacteria and viruses that have come to hatcheries and farms and will continue to reach them worldwide. Future experiments are being planned at CIBNOR to elucidate the role that homoeopathic medications could play in these organisms.

Acknowledgements

This study was financed by the Sectorial Fund for Education of Mexico: Project CB SEP-CONACYT-258282 "Experimental evaluation of homeopathy and new probiotics in the culture of mollusks, crustaceans and fish of commercial interest" and Project PROINNOVA-CONACYT-241777 "Innovation and continuous improvement of products and processes to optimise hatchery seed production of the American oyster *C. virginica*", under the academic responsibility of the first author (JMMS); the authors thank Diana Fischer for editorial services.

Author details

Mazón-Suástegui José Manuel[1], Tovar-Ramírez Dariel[1], Salas-Leiva Joan Sebastian[1]*, Arcos-Ortega Guadalupe Fabiola[1], García-Bernal Milagro[2], Avilés-Quevedo María Araceli[3], López-Carvallo Jesús Antonio[1], García-Corona José Luis[1], Ibarra-García Laura Elizabeth[1], Ortíz-Cornejo Nadia Livia[1], Teles Andressa[1], Rosero-García Antonia[4], Abasolo-Pacheco Fernando[5], Campa-Córdova Ángel Isidro[1], Saucedo Pedro[1], Barajas-Frías José Delfino[1], Ormart-Castro Pablo[1], Rodríguez-Jaramillo Carmen[1], González-González Reinaldo[1], Barajas-Ponce Ulysses[6], Tordecillas-Guillén Jorge Luis[6], Álvarez-Gil Francisco Antonio[7], Pineda-Mahr Gustavo[8], Peiro-López Jesús[8] and Robles-Mungaray Miguel[9]

*Address all correspondence to: jsalasleiva@gmail.com

1 Centro de Investigaciones Biológicas del Noroeste S.C., La Paz, B.C.S., México

2 Universidad Central de Las Villas (CBQ), Santa Clara, Villa Clara, Cuba

3 Freelance Services for R&D in Marine Aquaculture, La Paz, B.C.S., México

4 Centro Interdisciplinario de Ciencias Marinas I.P.N., La Paz, B.C.S., México

5 Universidad Técnica Estatal de Quevedo (FCA), Quevedo, Ecuador

6 Centro Ostrícola Tecnológico de Tabasco (Commercial Oyster Hatchery), Villahermosa, Tabasco, México

7 BCS Camarón (Commercial Shrimp Farm), La Paz, B.C.S., México

8 Acuacultura Mahr (Commercial Shrimp Hatchery), La Paz, B.C.S., México

9 Acuacultura Robles (Commercial Oyster Hatchery), La Paz, B.C.S., México

References

[1] Harikrishnan R, Balasundaram C, Heo M-S. Impact of plant products on innate and adaptive immune system of cultured finfish and shellfish. Aquaculture. 2011;**317**(1):1-15

[2] Romero J, Feijoó CG, Navarrete P. Antibiotics in aquaculture–Use, abuse and alternatives. In: Health and Environment in Aquaculture. Rijeka, Croatia: InTech; 2012

[3] Bernal MG, Marrero RM, Campa-Córdova ÁI, Mazón-Suástegui JM. Probiotic effect of Streptomyces strains alone or in combination with *Bacillus* and *Lactobacillus* in juveniles of the white shrimp *Litopenaeus vannamei*. Aquaculture International. 2017;**25**(2):927-939

[4] Kumar M, Kumar V, Roy D, Kushwaha R, Vaiswani S. Application of herbal feed additives in animal nutrition-a review. International Journal of Life Sciences. 2014;**4**:1-8

[5] Ortiz-Cornejo NL, Tovar-Ramírez D, Abasolo-Pacheco F, Mazón-Suástegui JM. Homeopatía, una alternativa para la acuicultura. Revista Médica de Homeopatía. 2017; **10**(1):18-24

[6] Toledo M, Stangarlin J, Bonato C. Homeopathy for the control of plant pathogens. Physiology. 2011;**19**(20):21

[7] Bellavite P, Marzotto M, Olioso D, Moratti E, Conforti A. High-dilution effects revisited. 1. Physicochemical aspects. Homeopathy. 2014;**103**(1):4-21

[8] Chikramane PS, Suresh AK, Bellare JR, Kane SG. Extreme homeopathic dilutions retain starting materials: A nanoparticulate perspective. Homeopathy. 2010;**99**(4):231-242

[9] Bell IR, Schwartz GE. Enhancement of adaptive biological effects by nanotechnology preparation methods in homeopathic medicines. Homeopathy. 2015;**104**(2):123-138

[10] Bellavite P, Ortolani R, Conforti A. Immunology and homeopathy. 3. Experimental studies on animal models. Evidence-Based Complementary and Alternative Medicine. 2006;**3**(2):171-186

[11] Andretto AP, Fuzinatto MM, Bonafe EG, Braccini GL, Mori RH, Pereira RR, et al. Effect of an homeopathic complex on fatty acids in muscle and performance of the Nile tilapia (*Oreochromis niloticus*). Homeopathy. 2014;**103**(3):178-185

[12] de Oliveira Feitosa KC, Povh JA, de Abreu JS. Physiological responses of pacu (*Piaractus mesopotamicus*) treated with homeopathic product and submitted to transport stress. Homeopathy. 2013;**102**(4):268-273

[13] Valentim-Zabott M, Vargas L, Ribeiro RPR, Piau Jr R, Torres MBA, Rönnau M, et al. Effects of a homeopathic complex in Nile tilapia (*Oreochromis niloticus L.*) on performance, sexual proportion and histology. Homeopathy. 2008;**97**(4):190-195

[14] Kesarcodi-Watson A, Kaspar H, Lategan MJ, Gibson L. Probiotics in aquaculture: The need, principles and mechanisms of action and screening processes. Aquaculture. 2008; **274**(1):1-14

[15] López-Carvallo JA, Mazón-Suástegui JM, Arcos-Ortega GF, Tovar-Ramírez D, Hernández-Oñate MA A-PF. Effect of homeopathic treatments on immune response in Catarina scallop (*Argopecten ventricosus*) spat. 2017;(Larvi-7th fish & Shellfish Larviculture Symposium). pp. 266-269

[16] Song L, Wang L, Qiu L, Zhang, H. Bivalve immunity. In Invertebrate immunity. USA: Springer; 2010. pp. 44-65

[17] Abasolo-Pacheco F, Campa-Córdova ÁI, Mazón-Suástegui JM, Tovar-Ramírez D, Araya R, Saucedo PE. Enhancing growth and resistance to *Vibrio alginolyticus* disease in catarina scallop (*Argopecten ventricosus*) with *Bacillus* and *Lactobacillus* probiotic strains during early development. Aquaculture Research. 2017;**48**(9):4597-4607

[18] Wang W, Sun J, Liu C, Xue Z. Application of immunostimulants in aquaculture: Current knowledge and future perspectives. Aquaculture Research. 2017;**48**(1):1-23

[19] Finney DJ. Probit Analysis 1971. New York, Ny. 1971;**10022**:32

[20] López-Carvallo JA, Saucedo PE, Rodríguez-Jaramillo C, órdova ÁIC-C, García-Corona JL, Mazón-Suástegui JM. Carbohydrate-rich diets meet energy demands of gametogenesis in hatchery-conditioned mussels (*Modiolus capax*) at increasing temperatures. Journal of Shellfish Research. 2017;**36**(3):649-657

[21] Puinean AM, Labadie P, Hill EM, Osada M, Kishida M, Nakao R, et al. Laboratory exposure to 17β-estradiol fails to induce vitellogenin and estrogen receptor gene expression in the marine invertebrate *Mytilus edulis*. Aquatic Toxicology. 2006;**79**(4):376-383

[22] Rodríguez-Jaramillo C, Hurtado MA, Romero-Vivas E, Ramírez JL, Manzano M, Palacios E. Gonadal development and histochemistry of the tropical oyster, *Crassostrea corteziensis* (Hertlein, 1951) during an annual reproductive cycle. Journal of Shellfish Research. 2008;**27**(5):1129-1141

[23] Mazón-Suástegui JM, García-Bernal M, Saucedo PE, Campa-Córdova Á, Abasolo-Pacheco F. Homeopathy outperforms antibiotics treatment in juvenile scallop *Argopecten ventricosus*: Effects on growth, survival, and immune response. Homeopathy. 2017;**106**(1):18-26

[24] Barber BJ, Blake NJ. Reproductive physiology. In: Developments in Aquaculture and Fisheries Science. Netherlands: Elsevier; 2006. pp. 357-416

[25] Vidal EAG, Villanueva R, Andrade JP, Gleadall IG, Iglesias J, Koueta N, et al. Cephalopod culture: Current status of main biological models and research priorities. In: Advances in marine biology. Netherlands: Elsevier; 2014. p. 1-98

[26] Villanueva R, Sykes AV, Vidal EAG, Rosas C, Nabhitabhata J, Fuentes L, Iglesias J. Current status and future challenges in cephalopod culture. In: Iglesias J, Fuentes L, Villanueva R, editors. Cephalopod Culture. Netherlands: Springer; 2014. pp. 479-489

[27] Santé-Lhoutellier V, Astruc T, Marinova P, Greve E, Gatellier P. Effect of meat cooking on physicochemical state and in vitro digestibility of myofibrillar proteins. Journal of Agricultural and Food Chemistry. 2008;**56**(4):1488-1494

[28] Domingues PM, Marquez L, López N, Rosas C. Effects of food thermal treatment on growth, absorption, and assimilation efficiency of juvenile cuttlefish, *Sepia officinalis*. Aquaculture International. 2009;**17**(3):283

[29] Rosas C, Valero A, Caamal-Monsreal C, Uriarte I, Farias A, Gallardo P, et al. Effects of dietary protein sources on growth, survival and digestive capacity of *Octopus maya* juveniles (Mollusca: Cephalopoda). Aquaculture Research. 2013;**44**(7):1029-1044

[30] Rodríguez-González T, Valverde JC, Sykes AV, García BG. Performance of raw material thermal treatment on formulated feeds for common octopus (*Octopus vulgaris*) ongrowing. Aquaculture. 2015;**442**:37-43

[31] Vockeroth WG. Veterinary homeopathy: An overview. The Canadian Veterinary Journal. 1999;**40**(8):592

[32] Akazawa N, Eguchi M. Environmental trigger for EMS/AHPNS identified in Agrobest shrimp ponds. Global Aquaculture Advocate. 2013:16-17

[33] Doehring C, Sundrum A. Efficacy of homeopathy in livestock according to peer-reviewed publications from 1981 to 2014. The Veterinary Record. 2016;**179**(24):628

[34] Khalil SA, Abou-Akkada AS, El-Hoshy SM. Molecular studies on vibrio species isolated from imported frozen fish. Global Veterinaria. 2014;**12**(6):782-789

[35] Liu AX, Liao ZW. Effects of nano-materials on water clusters. Journal of Anhui Agricultural Sciences. 2008;**36**(36):15780-15781

Chapter 5

Gut Microbiota and Innate Immune Response of *Macrobrachium vollenhovenii* Infected with *Pseudomonas aeruginosa* and *Aeromonas hydrophila* Fed Diets Supplemented with *Lactobacillus acidophilus*

Ibrahim Adeshina, Yusuf Adetunji Adewale,
Lateef Oloyede Tiamiyu,
Mozeedah Modupeola Ajibola and
Akeem Babatunde Dauda

Additional information is available at the end of the chapter

http://dx.doi.org/10.5772/intechopen.78010

Abstract

The use of antibiotics as disease control agents has become contentious due to rise in drug-resistant bacteria such as *Psuedomonas aeruginosa* and *Aeromonas hydrophilla*. Studies have shown antibacterial potentials of some probiotics such as *Lactobacillus acidophillus* as promising alternative. Therefore, effects of diets fortified with *Lactobacillus acidophillus* on gut ecology and health status of African River prawn, *Macrobranchium vollenhovenii* were investigated. Prawns fed diets fortified with *Lactobacillus acidophillus* were challenged with *Psuedomonas aeruginosa* (1×10^7 cfu/mL) and *Aeromonas hydrophilla* (5×10^5 cfu/mL) using bath method for 14 days. Total viable and total enterobacteriaceae counts were determined on plate count agar and McConkey agar, respectively. Haemolymph (mL), total haemocyte count (cells/mL), catalase (mg/g protein), superoxide dismutase (mg/g protein, respiratory burst activity (μmoles) and survival rate (%) were evaluated using standard procedures. Data were analysed using descriptive statistics and ANOVA at $p = 0.05$. Results revealed that *Marobranchium vollenhovenii*-fed supplemented diets had reduced bacteria load, gut flora dominated by beneficial bacteria, enhanced immune system and protection against *Aeromonas hydrophila* at 10^3 cfu/mL inclusion level and could be used as immunodulation against *Aeromonas* infection.

Keywords: gut microbiota, innate immune response, infection, bacteria

1. Introduction

Prawn such as *Macrobrachium vollenhobenii* is an important source animal protein that provides good quality protein and essential minerals and vitamins. It is also regarded as safest animal source of protein which comes from either wild or culture system. The current decline in the status fish from wild has been attributed to overfishing, illegal and irresponsible fishing, habitat destruction and pollution, among other factors. More so, the increase in population and change in the consumption pattern of fish has create huge gap between the demand and supply of fish [1]. One of the ways to bridge this gap is through aquaculture. The act of aquaculture means rearing of aquatic organisms, which entails any form of intervention to improve the production quantity and quality, management, disease prevention and control. Over the years, aquaculture has become the fastest growing sector in the world [2].

Worldwide total crusteceans was 6,915,100 tonnes representing about 7% of total aquaculture production [2]. This achievement was attributed to transformation of farming techniques from simple methods to an improved and intensive aquaculture which promotes high seed quantities, adequate knowledge of stocking density, supply of required fish feeds. However, in spite of the success recorded in the prawn farms in recent times, the performance are far from it demand. The poor performance has been associated with many constraints, such as, modern knowledge in the science of fish farming, government policy, fish feed industry, marketing, distribution, and diseases. One of the major barriers to prawn farming are diseases and their management.

Disease is an establishment of pathogens in prawn tissues which cause disorderliness in physiological function of the fish that result in physical, biological and economical losses. Diseases arise as a result of complex interaction among the fish, pathogen and culture environment [3, 4]. Fish has inbuilt immune systems and defence mechanisms which protect them from being infected with pathogens. But practice of intensive farming system has the possibility of exposing prawn to infections. However, bacteria have been reported to be responsible for about 70–80% of disease infection in fish [5]. Some important bacteria in prawn farming are *Pseudomonas aeruginosa* and *Aeromonas hydrophila. Pseudomonas aeruginosa* and *Aeromonas hydrophila* are gram-negative bacteria in the families of Pseudomonadaceae and Aeromonadaceae respectively. They are ubiquitous, facultative anaerobe, rod-shaped and sugar fermented organism. Studies have reported that *Pseudomonas aeruginosa* and *Aeromonas hydrophila* infection in fish has resulted into haemorrhagic, speticemia, furunculosis and high mortality among others [1].

The problems of increase antibiotic resistance bacteria, residual effect and environmental unfriendly experienced in the use of antibiotics could ameliorate by probiotics application. Therefore, current studies have moved toward search for alternative such as probiotics. Probiotics are life microbial feed supplements that improve health host by modify the gastrointestinal tract of the fish. Fish, being a hydrophilic animal rely solely on the environment (water) which filtering through the body and gill as fish performs it physiological function would benefit from use of probiotics. Probiotics enhance the nutrient utilisation, modulate gut flora, inhibit the growth of pathogenic bacteria and improve growth and immune system of the fish as reported in the previous studies [6, 7]. Several probiotics have been used in

http://dx.doi.org/10.5772/intechopen.78010

aquaculture but probiotics from lactic acid bacteria (LAB) and *Bacillus* species are often used [7, 8].

Lactobacillus acidophilus is bacteria belonging to the genus Bacillus. It probiotics benefit as member of LAB group to improve non-specific immune response and disease resistance in in *Macrobrachium vollenhovenii* has not been fully elucidated hence the need for this study. Therefore, this study investigated the effect of diets fortified with *Lactobacillus acidophilus* on gut microbiota and innate immune response of *Macrobrachium vollenhovenii* Infected with *Pseudomonas aeruginosa* and *Aeromonas hydrophila.*

2. Materials and methods

2.1. Preparation of experimental diets

Feed ingredients were purchased from a reliable store Melbourne, Australia. *Lactobacillus acidophilus* ATCC 4356 was obtained from a laboratory and prepared into 0, 10^1, 10^2, 10^3, 10^4 and 10^5 cfu/mL (**Table 1**). Soya bean was further prepared by toasting in a hot plate for 15 min at 100°C. Ingredients were ground in hammer mill and mixed together to formulate 43% crude protein (**Table 1**) calculated according to Pearson Square Method [9]. The mixed ingredients were pelleted through pelleting machine. The feeds were packed in polythene bags and stored in a cool dry place at room temperature until use with labels on them. The crude proteins of the ingredients were: fish meal (72%); toasted soybean (46.2%) and white maize (9.3%) [10]. The feeds were reproduced after 21 days to avoid nutrients depletion.

2.2. Experimental design and procedure

Macrobrachium vollenhenvonii juveniles (mean weight = 18.02 ± 0.11 g; n = 360; 180 prawns for each pathogen) were obtained from a reputable farm in Melbourne, Australia and acclimatised for 2 weeks in glass aquaria tanks before the experiment. Prawns were weighed and distributed into 18 glass aquaria tanks (35 × 30 × 20 cm^3) in a completely randomised design with three replicates. Each tank contained 20 prawns. The tanks were constantly connected to aerator (Model: AP-60) with air blowers. The prawns that were fed experimental diets were further subjected to *Pseudomonas aeruginosa* and *Aeromonas hydrophila* infection for 2 weeks to examine their innate immune performances and survival. The diets were fed to the *Macrobrachium vollenhovenii* to satiation throughout the experimentation. Measurements of the weight changes were measured using sensitive scale (**Model: M1207**).

2.3. Evaluation of gut microbiota

Three prawns were collected from each experimental unit before and after the trial for gut microbiota evaluation. Each prawn was deactivated in freezer at −20°C for 10 min and sterilised using formalin (50 ppm). The guts were aseptically collected and weighed into sterile universal bottles containing peptone water (0.1%) to release the available bacteria for a period of 2 h. 1 mL was taken from each sample bottle and diluted 10-folds and subsequently serially

Ingredients (%)	*Lactobacillus acidophilus* inclusion levels cfu/mL					
Groundnut cake	25.00	25.00	25.00	25.00	25.00	25.00
Soya meal	24.00	24.00	24.00	24.00	24.00	24.00
Fish meal	22.15	22.15	22.15	22.15	22.15	22.15
DCP	1.00	1.00	1.00	1.00	1.00	1.00
Salt	1.00	1.00	1.00	1.00	1.00	1.00
Flour	0.89	0.89	0.89	0.89	0.89	0.89
Vegetable oil	0.50	0.50	0.50	0.50	0.50	0.50
Maize	24.96	24.96	24.96	24.96	24.96	24.96
Premix*	0.50	0.50	0.50	0.50	0.50	0.50
LA	0	10^1	10^2	10^3	10^4	10^5
TOTAL	100.00	100.00	100.00	100.00	100.00	100.00
Proximate composition						
Moisture	9.16	9.16	9.16	9.16	9.16	9.16
Crude protein	40.00	40.04	40.10	40.11	40.12	40.12
Ether extract	9.63	9.63	9.54	9.63	9.66	9.65
Total ash	8.27	8.27	8.27	8.27	8.27	8.27
Carbohydrates	20.17	20.13	20.16	20.06	20.02	20.03
Crude fibre	12.77	12.77	12.77	12.77	12.77	12.77

Note: DCP = Dicalcium phosphate; LA = *Lactobacillus acidophilus*.*Premixes = HI-MIX®AQUA (Fish) each 1 kg contains; vitamin A, 4000,000 International Unit (IU); vitamin D3, 8,00,000 IU; vitamin E, 40, 000 IU; vitamin K3, 1600 mg; vitamin B1, 4000 mg; vitamin B2, 3000 mg; vitamin B6, 3800 mg; vitamin B12, 3 mcg; Nicotinic acid 18,000 mg; Pantothenic acid, 8000 mg; Folic acid, 800 mg; Biotin, 100 mcg; Choline chloride 120,000 mg; Iron, 8000 mg; Copper, 800 mg; Manganese, 6000 mg; Zinc, 20,000 mg; Iodine, 400 mg; Selenium, 40 mg; Vitamin C C(coated), 60,000 mg; Inositol, 10,000 mg; Colbat, 150 mg; Lysine, 10,000 mg; Methionine, 10,000 mg; Antioxidant, 25,000 mg.

Table 1. Ingredients and chemical composition (%; on dry matter basis) of experimental diets containing graded levels of *Lactobacillus acidophilus*.

diluted with dilution factor of 10^{-4}. 2 mL was taken from each diluted sample and dispensed into two Petri dishes (1 mL to each). The first dish received plate count agar (PCA, LAB M, LAB149) for total viable count (TVC), while the second Petri dish received MacConkey agar (LAB M, LAB002) for total enterobactereceae count (TEB) using the pure plate count method [11–12]. Each dilution was overlaid, respectively, with PCA and MacConkey that have been cooled to 50°C. At this temperature, agar is still in liquid form [12].

The dishes were then gently swirled to mix the bacteria with the liquid agar. The mixtures were allowed to harden. When the mixture was hardened, the individual cells were fixed in place and incubated (Newlife Laboratory Incubator NL-9052-1) for 24 h at 37°C to allow distinguished colonies to form. The colonies formed were counted using Wincom Colony Counter (16 W, 220 V ± 10%, 50 Hz). The experiments were replicated three times. The TVC and TEB were expressed in Log_{10}CFU/g [12].

2.4. Bacterial challenge

To evaluate prawn resistance to disease infection, a challenge test using the pathogenic bacteria, *Pseudomonas aeruginosa* and *Aeromonas hydrophila* was performed following 84 days feeding trial. Prior to the challenge test, a preliminary experiment was performed to determine the LD_{50} (lethal dose) of the pathogenic bacteria. For the challenge test, 10 prawn were selected from each replicate tank and transferred into another tank filled with dechlorinated freshwater. *Pseudomonas aeruginosa* was grown nutrient agar broth and *Aeromonas hydrophila* on PBS broth for 24 h at 35°C in an incubator in Microbiology Laboratory. Bacterial cells were then centrifuged at 3000× g for 30 min to form pellets. The pellets were re-suspended in 1 mL of 0.1% peptone water and after which 100 mL of the suspension were diluted at 4:10 with water from experimental tank. The prawn were exposed to bacterial solution of both *Pseudomonas aeruginosa* (1×10^7 CFU/mL) and *Aeromonas hydrophila* (5×10^5 CFU/mL) for 30 min and returned into experimental set up [13, 14]. The fish were fasted for 24 before infection and feeding with experimental diets resumed 12 h later. All prawn groups were kept under observation for 14 days to record any abnormal clinical signs and the daily mortality.

2.5. Determination of innate immune response parameters

Three prawns were randomly selected from each experimental unit. The haemolymph was collected from the ventral part of the haemocoel of the second abdominal segment with the aid of a sterilised syringe and a 21-gauge disposable hypodermic needle containing 1 mL of Alserver's solution and was transferred into anticoagulant bottle (EDTA). The plasma was prepared by centrifuged the haemolymph at 300× g for 10 min at 4°C. The haemocytes were suspended and adjusted to a concentration of 5×10^6 cells/mL in an ice-cold.

The innate immune parameters were measured using the diagnostic reagent kits (Randox® Laboratories, Crumlin, County Antrim, UK). Superoxide dismutase (SOD) activity was measured spectrophotochemically by the ferricytochrome c method using xanthine/xanthine oxidase as the source of superoxide radicals. The reaction mixture consisted of 50 mM potassium phosphate buffer (pH 7.8), 0.1 mM EDTA, 0.1 mM xanthine, 0.013 mM cytochrome C and 0.024 IU/mL xanthine oxidase. One activity unit was defined as the amount of enzyme necessary to produce a 50% inhibition of the ferricytochrome C reduction rate measured at 550 nm. Catalase (CAT) activity was determined by measuring the decrease of H_2O_2 concentration at 240 nm according to [15]. The reaction mixture contained 50 mm potassium phosphate buffer (pH 7.0) and 10.6 mM H_2O_2 freshly prepared.

The respiratory burst activity was measured using diagnostic reagent kits (Randox, London, UK) as described by [16]. Respiratory burst activity was quantified by the nitroblue tetrazolium (NBT) assay which measures the quantity of intracellular oxidative free radicals; according to [17], with some modification. Briefly, 100 mL of the haemocytes were added to each well of a 96 well microtitre plate (Nalge-Nunc, Hereford, UK). The plate was incubated at 25°C, for 2 h to allow attachment of cells. Unattached cells were washed off three times using fresh L-15 medium. L-15 medium was then supplemented with NBT (1 mg/mL) and phorbol 12-myristate 13-acetate (PMA, SigmaeAldrich; 1 mg/mL) dissolved in dimethyl sulphoxide (DMSO, Sigma), and 100 mL added to each well of the microtitre plate and incubated for 1 h at room temperature. After incubation, the supernatant removed from the plate and NBT reduction fixed with 100% methanol for 10 min. The plate was then washed with 70% methanol, and left to air dry. A mixture of 120 mL of 2 M potassium hydroxide and 140 mL DMSO was added to dissolve the resulting formazan blue crystals. The NBT reduction was measured using the microplate reader (Optica, Mikura Ltd., UK) at 630 nm, and respiratory burst activity was expressed as NBT reduction.

Total haemocyte count (THC) was performed in a haemocytomter using microscope. The phenolosidase activity (PO) was evaluated by measuring the formation of dopachrome L-dihydrophenylclamine (L-DOPA) at 490 nm with the aid of spectrophotometer. While reactive oxygen intermediates (ROI) were used to measure H_2O_2 by horseadish peroxidase (dependent oxidation of phenol red) while chemiluminescence was used to measure the light emission from reactive oxygen intermediates [18]. Lysozyme activity of fish sera was determined by using lysoplate technique [19]. In brief, 0.60 mg/mL *Micrococcus luteus* was cast in 1% agarose gel (Difco, USA) with 50 mM phosphate buffer (pH 6.2). Wells (6 mm) were created nutrient agar plates and were filled 25 μL of serum samples and incubated for 20 h at 25°C. Lysozyme activity was calculated from a standard curve prepared with lysozyme from chicken egg white. The respiratory burst activity was measured using diagnostic reagent kits (Randox, London, UK) as described by Chiu et al. (2007). Relative protection level (RPL) was estimated as RPL = [(1−%mortality in treatment)/% mortality in control) × 100] [20].

2.6. Statistical analysis

The results were presented as mean ± SE of three replicates. Prior to statistical analysis, all data were tested for normality of distribution using the Kolmogorov–Smirnov test. The homogeneity of variances among different treatments was tested using Bartlett's test. Then, data were subjected to one-way ANOVA to evaluate effects of *Lactobacillus acidophilus* supplementation. Differences between means were tested at the 5% probability level using Duncan test. The optimum *Lactobacillus acidophilus* level was determined using polynomial regression analysis. All the statistical analyses were done using SPSS program version 20 (SPSS, Richmond, VA, USA).

3. Results

The effects of *Lactobacillus acidophilus* as a supplement on gut microbiota of *Macrobranchium vollenhovenii* are presented in **Table 2**. The total viable bacteria count (TVC) was significantly decreased in the gut samples of prawn fed with feed supplemented diets ($p < 0.05$). Similarly, total enterobactereceae (TEB) was reduced significantly in fish fed fortified diets. The highest TVC and TEB were recorded in fish fed control diet.

Table 3 depicts microbiota composition of *Macrobranchium vollenhevenii* fed diets fortified with *Lactobacillus acidophilus*. The results indicated the species composition of the prawn varied significantly with respect to levels of inclusion of *Lactobacillus acidophilus*. The dominated species were *Pseudomonas aeruginosa, Aeromonas hydrophila, Bacillus and Staphylococcus* species.

***Lactobacillus acidophilus* inclusion levels (cfu/mL)**	**Parameters (log_{10}cfu/mL)**	
	TVC[*]	**TEB[*]**
Control	6.32 ± 0.03[a]	6.12 ± 0.02
10^1	6.21 ± 0.04	5.96 ± 0.11
10^2	6.18 ± 0.20	5.72 ± 0.38
10^3	6.01 ± 0.16	5.62 ± 0.12
10^4	5.91 ± 0.09	5.43 ± 0.13[a]
10^5	5.81 ± 0.45	5.40 ± 0.90[a]

Means (Log_{10} cgu/mL) with different superscripts are significantly different ($p < 0.05$), while, absence of letters means no significantly different ($P > 0.05$). TVC = Total viable bacteria count; TEB = Total enterobacterceae.

Table 2. Gut microbiota of *Macrobranchium vollenhovenii* fed diets fortified with *Lactobacillus acidophilus*.

Parameters (log_{10}cfu/mL)	***Lactobacillus acidophilus* inclusion levels (cfu/mL)**					
	Control	**10^1**	**10^2**	**10^3**	**10^4**	**10^5**
Pseudonomas aeruginosa	6.02 ± 0.09	5.92 ± 0.06	5.63 ± 0.26	5.57 ± 0.28a	5.32 ± 0.04a	5.15 ± 0.32a
Aeromonas hydrophila	6.06 ± 0.29a	5.49 ± 0.03a	5.38 ± 0.07	5.20 ± 0.22	4.90 ± 0.19	4.85 ± 0.04
Bacillus species	4.98 ± 0.07a	6.08 ± 0.58	6.14 ± 0.02	6.15 ± 0.03	6.17 ± 0.41	6.26 ± 0.21
Staphylococcus species	6.24 ± 0.03	6.22 ± 0.91	6.20 ± 0.15	6.19 ± 0.33	6.18 ± 0.21	6.18 ± 0.05

Means (Log_{10} cgu/mL) with different superscripts are significantly different ($p < 0.05$), while, absence of letters means no significantly different ($P > 0.05$).

Table 3. *Gut microbiota composition of Macrobranchium vollenhovenii fed diets fortified with lactobacillus acidophilus.*

Pseudomonasa aeruginosa and Aeromonas hydrophila were reduced significantly ($p < 0.05$) while *Bacillus* species was greatly increased. However, there was no significantly difference in the load of *Staphylococcus* species. The highest *Pseudomonas aeruginosa, Aeromonas hydrophila* and *Staphylococcus* species were recorded in prawn fed control diet while the least was observed in prawn fed diet fortified with 10^5 cfu/mL *Lactobacilus acidophilus*. However, the highest *Bacillus* species was obtained in prawn fed diet supplemented with 10^5 cfu/mL *Lactobacilus acidophilus*.

Effects of *Lactobacillus acidophilus* based diets on innate immune responses of *Macrobranchium vollenhovenii* was presented in **Table 4**. The immune response of the prawn was stimulated by *Lactobacillus acidophilus* supplementation. The SOD, CAT, RBA, THC, PO, ROI and lysozyme activity were significantly different ($p < 0.05$) among the treatments. Highest SOD, CAT, RBA, THC, PO, ROI and lysozyme activity were recorded in prawn fed 10^5 cfu/mL diet; meanwhile, the lowest values were obtained in the control group.

Parameters (log_{10}cfu/mL)	*Lactobacillus acidophilus* inclusion levels (cfu/mL)					
	Control	10^1	10^2	10^3	10^4	10^5
SOD (mg/g protein)	0.41 ± 0.01a	2.45 ± 0.02ab	3.62 ± 0.05b	6.33 ± 0.15	6.57 ± 0.55	6.82 ± 0.22
CAT mg/g protein)	1.51 ± 0.02a	2.04 ± 0.04a	2.74 ± 0.01	3.06 ± 0.20	3.47 ± 0.21	3.85 ± 0.04
RBA	165.4 ± 2.17a	163.1 ± 3.13a	172.8 ± 1.11b	188.0 ± 2.15b	215.4 ± 2.46	243.7 ± 2.85
THC	12.06 ± 0.26a	13.11 ± 0.05	13.53 ± 0.18	14.02 ± 0.17	14.65 ± 0.78	14.82 ± 0.16
PO (U/mL)	3.24 ± 0.43a	3.51 ± 0.07	4.07 ± 0.16	4.33 ± 0.09	4.89 ± 0.31	5.21 ± 0.34
ROI	5.27 ± 0.21a	6.02 ± 0.15	6.31 ± 0.34	6.38 ± 0.21	6.54 ± 0.23	6.79 ± 0.29
Lysozyme activity	8.13 ± 0.62a	11.20 ± 0.38a	13.05 ± 0.03	13.46 ± 0.78	13.78 ± 0.19	14.02 ± 0.18

Means with different superscripts are significantly different ($p < 0.05$), while, absence of letters means no significantly different ($P > 0.05$).

Table 4. Innate immune response parameters of *Macrobranchium vollenhovenii* fed diets fortified with *Lactobacillus acidophilus*.

In addition, the exposure of prawn fed fortified diets to *Pseudomonas aeruginosa* infection was presented in **Table 5**. The result indicated that there was significant difference ($p < 0.05$) in survival rate and relative protection of prawn fed fortified diets. Highest survival rate and relative protection were recorded in group fed 10^5 cfu/mL *Lactobacilus acidophilus* inclusion level and least were recorded in control group.

Table 6 reveals the resistant of *Macrobranchium vollenhovenii* to *Aeromonas hydrophila* infection. Survival and relative protection of prawn to *Aeromonas hydrophila* were significantly different ($p < 0.05$). Prawn fed 10^5 cfu/mL *Lactobacilus acidophilus* inclusion level had highest survival and relative protection rates while the prawn treated control diet had the lowest survival and relative protection rates.

Lactobacillus acidophilus inclusion levels (cfu/mL)	Parameters (log_{10}cfu/mL)		
	Initial stock	Survival (%)	Relative protection
Control	30	23.3 ± 0.03a	0.00 ± 0.00a
10^1	30	40.0 ± 0.13b	21.74 ± 0.15b
10^2	30	73.3 ± 0.23c	65.22 ± 0.16c
10^3	30	86.7 ± 0.71	82.61 ± 0.37
10^4	30	93.3 ± 0.11	91.30 ± 0.21
10^5	30	96.7 ± 0.42	95.65 ± 0.02

Means with different superscripts are significantly different ($p < 0.05$), while, absence of letters means no significantly different ($P > 0.05$).

Table 5. Resistant of *Macrobranchium vollenhovenii* fed diets fortified with *Lactobacillus acidophilus* to *Pseudomonas aeruginosa* infection.

***Lactobacillus acidophilus* inclusion levels (cfu/mL)**	**Parameters (log_{10}cfu/mL)**		
	Initial stock	**Survival (%)**	**Relative protection**
Control	30	10.0 ± 0.01a	0.00 ± 0.00a
10^1	30	30.0 ± 0.01b	22.22 ± 0.03b
10^2	30	36.7 ± 0.15b	29.63 ± 0.11b
10^3	30	83.3 ± 0.21	81.48 ± 0.16
10^4	30	90.0 ± 0.13	88.89 ± 0.80
10^5	30	93.3 ± 0.04	92.59 ± 0.51

Means with different superscripts are significantly different ($p < 0.05$), while, absence of letters means no significantly different ($P > 0.05$).

Table 6. Resistant of *Macrobranchium vollenhovenii* fed diets fortified with *Lactobacillus acidophilus* to *Aeromonas hydrophila* infection.

4. Discussion

The result of the gut flora revealed that there were reductions in the bacteria load of the prawn fed fortified diets. The decrease in the in observed could be attributed to the activities of *Lactobacillus acidophilus* as probiotics which modulate the gut pH to its favour and outcompete the pathogenic organisms. According to International Commission on the Microbiological Specification of Foods [21] the acceptable level of bacterial load in fish tissue should be lower than 5.70 Log_{10}cfu/g. In this study, the TVC and TEB of prawn fed fortified diets were within the recommended values except the control and fish fed 10^1 cfu/mL of *Lactobacillus acidophilus*. The reduction in the load of bacteria especially TEB signifies that the probiotics improve the gut of the prawn by outcompete the pathogenic bacteria. The gut flora comprises *Pseudomonas aeruginosa, Aeromonas hydrophila,* Bacillus species and *Staphylococcus* species. Fish fed *Lactobacillus acidophilus* based diets had higher gut flora dominated by *Bacillus* species that the fish fed control diet and reduction of pathogenic organisms were observed. The findings of this study are in agreement with the work of [22] who reported that probiotics enhanced the population of beneficial bacteria and suppressed the growth of pathogenic bacteria.

Gut flora have continuous and dynamic effect on the host's gut and systemic immune systems. The bacteria are key in promoting the early development of the gut's mucosal immune system both in terms of its physical components and function. The bacteria stimulate the lymphoid tissue associated with the gut mucosa to produce antibodies to pathogens [23]. The immune system recognises and fights harmful bacteria, but leaves the helpful species alone, tolerance developed in juveniles [24]. Recent findings have shown that gut bacteria play a role in the expression of toll-like receptors (TLRs) in the intestines, molecules that help the host repair damage due to injury. The TLRs are one of the two classes of pattern-recognition receptors (PRR) that provide the intestine the ability to discriminate between the pathogenic and commensal bacteria [25]. These PRRs identify the pathogens that have crossed the mucosal barriers and trigger a set of responses that take action against the pathogen [26].

Also, [27] claims that different species of gut flora could influence the development of key cells of the immune system, by increasing or decreasing the level of tolerance against foreign entities. Once the host immune system is developed, it regulates the bacterial composition in the gut. One of the regulation mechanisms—immune exclusion—is mediated through the neutralisation of secreted immunoglobulin A (IgA) [28].

Harmful bacteria species, such as *Pseudomonas aeruginosa* and *Aeromonas hydrophila,* are unable to grow excessively due to competition from helpful gut flora species adhering to the mucosal lining of the intestine; thus, fish without/with low gut flora are infected very easily [28]. The barrier effect protects hosts from both invading species and species normally present in the gut at low numbers, whose growth is usually inhibited by the gut flora [29]. Helpful bacteria prevent the growth of pathogenic species by competing for nutrients and attachment sites to the epithelium of the colon. Also, symbiotic bacteria are more at home in this ecological niche and are thus more successful in the competition. Probiotics also produce bacteriocins, which are proteinaceous toxins that inhibit growth of similar bacterial strains, substances that kill harmful microbes and the levels of which can be regulated by enzymes produced by the host. Also, the process of fermentation produces lactic acid and different fatty acids, which lowers the pH in the colon, preventing the proliferation of harmful species of bacteria and facilitating that of helpful species [30].

Immune responses of the prawn fed fortified diets were higher than the control. There higher SOD, CAT RBA, THC, PO, ROI and lysozyme activity in prawn fed the *Lactobacillus acidophilus*-based diets than in those fed control diet. Similar observations were made by [22, 31] in fish fed probiotics diets. The SOD, CAT RBA, THC, PO, ROI and lysozyme activity have important roles in the innate immune defence system. For instance, it was reported that lysozyme activity plays a significant role in innate humoral immune factors that control intrusiveness of pathogens by destroying the cell walls of the bacteria [1]. The SOD catalyses the dismutation of the superoxide anions to molecular oxygen and hydrogen peroxide, which further catalyses to water and oxygen. The SOD and CAT play an important role as cellular antioxidants against reactive oxygen species. In this study, prawn fed fortified diets had increased values of SOD, CAT, RBA and lysozyme activity, which could be answerable for the increased concentration of hydrogen peroxide in the oxidation process. The chemicals are released from immune cells and enzymes, as they come in contact with pathogens cumulating into improved immunity.

The investigation into the resistance level of *Macrobranchium vollenhovenii* fed diets fortified with *Lactobacillus acidophilus* revealed that the prawn were protected that the group fed control diet. In this study, prawns were significantly protected against *Pseudomonas aeruginosa* and *Aeropmonas hydrophila* infection with relative protection up to 96%. Similarly, [32–33, 7] reported that there was significant increase in the survival rate of fish fed *Lactobacillus acidophilus* fortified diets. The protection of prawn against the pathogens could be attributed to the gut serves as an entry point through feeding and thus inhibit the growth and survival of the pathogens.

5. Conclusion

Macrobranhium vollenhovenii fed diets supplemented with *Lactobacillus acidophilus* had reduced total enterobactericeae in the gut, enhanced innate immune response parameters and suggest it usage as immunomodulation and protective agent against *Pseudomonas aeruginosa* and *Aeromonas hydrophila* infection.

Author details

Ibrahim Adeshina[1]*, Yusuf Adetunji Adewale[2], Lateef Oloyede Tiamiyu[1], Mozeedah Modupeola Ajibola[3] and Akeem Babatunde Dauda[4]

*Address all correspondence to: adesina.i@unilorin.edu.ng

1 Department of Aquaculture and Fisheries, University of Ilorin, Nigeria

2 School of Environmental Life Sciences, Faculty of Science and Built Engineering, Deakin University, Australia

3 Department of Zoology, University of Lagos, Nigeria

4 Department of Fisheries and Aquaculture, Federal University Dut-sima, Nigeria

References

[1] Adeshina I, Adewale YA, Tiamiyu LO. Growth performance and innate immune response of *Clarias gariepinus* infected with *Aeromonas hydrophila* fed diets fortified with *Curcuma longa* leaf. West Africa Journal of Applied Ecology. 2007;**25**(2017):79-90

[2] FAO (Food and Agriculture Organisation of the United Nations). The State of World Fisheries and Aquaculture: Contributing to Food Security and Nutrition for all. Rome Italy: FAO; 2016. p. 24

[3] Adedeji OB, Okocha RC. Constraint to aquaculture development in Nigeria and way forward. Journal of Applied Science and Research. 2011;**7**:1133-1140

[4] Ajani EK, Akinwole AO, Ayodele IA. Fundamentals of Fish Farming in Nigeria. 1st ed. Vol. 158. Ibadan: Walecrowns Ventures Nigeria; 2011

[5] Jami M, Ghanbari M, Zunabovic M, Domig KJ, Kneifel W. *Listeria monocytogenes* in aquatic food products–A review. Comparative Review of Food Science and Food Safety. 2014;**13**:798-813

[6] Nayak SK. Probiotics and immunity: A fish perspective. Fish and Shellfish Immunology. 2010;**29**:2-14

[7] Ulukoy G, Metin S, Kubilay A, Guney S, Yilidrim P, Guzel-Seydim Z, Kok-Tas T, Erkan G. The effect of kefir as a dietary supplement on nonspecific immune response and disease resistance in juvenile rainbow trout, *oncorhynchus mykiss* (Walbaum 1792). Journal of the World Aquaculture Society. 2017;**48**:248-256

[8] Munoz-Atienza E, Gomez-Sala B, Araujo C, Campanero C, Del Campo R, Hernandez PE. Antimicrobial activity, antibiotic susceptibility and virulence factor of lactic acid bacteria of aquatic origin intended for use as probiotics in aquaculture. BMC Microbiology. 2013;**24**:13-15

[9] FAO (Food and Agriculture Organisation of United Nation). Protein Quality Evaluation (Report of Joint FAO/WHO Constitution) Held in Bethside, M.D. USA, DE: Taylor and Franscis; 1990. pp. 10-35

[10] AOAC (Association of Official Analytical Chemist). Official Methods of Analysis. 18th ed. Maryland, USA: Association of Official Analytical, Chemists International; 2005

[11] Adeshina I, Jenyo-Oni A, Emikpe BO, Ajani EK. Effect of solvents on phytoconstituents and antimicrobial activities of *Ocimum gratissimum* and *Eugenia caryophyllata* extracts on Listeria monocytogenes. Acta Veterinarian Eurasia. 2018;**44**:31-38

[12] APHA (American Public Health Association). Standard Methods for the Examination of Bacteria in Water and Wastewater. 16th ed. Washington: American Public Health Association, American Water Works Association and Water Pollution Control Federation; 1985

[13] Ikeh MAC, Obi SKC, Ezeasor DN, Ezeonu IM, Moneke AN. Incidence and pathogenicity profile of Listeria sp. isolated from food and environmental samples in Nsukka. Nigeria. African Journal of Biotechnology. 2010;**9**:4776-4782

[14] Abdel-Tawwab M, Abbass FE. Turmeric powder (*Curcuma longa* L.) in common carp, *Cyprinus carpio* L., diets: Growth performance, innate immunity, and challenge against pathogenic *Aeromonas hydrophila* infection. Journal of the World Aquaculture Society. 2017;**48**:303-312

[15] Aebi H. Catalase in vitro. Methods in Enzymology. 1984;**105**:121-126

[16] Chiu CH, Guu YK, Liu CH, Pan TM, Cheng W. Immune responses and gene expression in white shrimp, *Litopenaeus vannamei,* induced by *Lactobacillus plantarum.* Fish and Shelfish Immunology. 2007;**23**:364-377

[17] Secombes CJ. Enhancement of fish phagocyte activity. Fish and Shellfish Immunology. 1994;**4**:421-436

[18] Anderson DP. Fish Immunology. 2nd ed. Hong Kong: TFH Publications, Inc. Ltd; 1977. p. 150

[19] Grinde B, Jolles J, Jolles P. Purification and characterization of two lysozymes from rainbow trout (*Salmo gairdneri*). European Journal of Biochemistry. 1988;**173**:269-273

[20] Amend DF. Potency testing of fish vaccines. Developments in Biologival Standardization. 1981;**49**:447-429

[21] ICMSF (International Commission on the Microbiological Specification of Foods). Sampling for Microbiological Analysis, Principles and Specific Applications of Microorganisms in Food. 2nd ed. Canada: Toronto: University of Toronto Press; 1982

[22] Addo S, Carrias AA, Williams MA, Liles MR, Terhune JS, Davis DA. Effects of *Bacillus subtilis* strain on growth, immune parameters, and *Streptococcus iniae* susceptibility in Nile Tilapia, *Orechromis niloticus*. Journal of the World Aquaculture Society. 2017;**48**:257-267

[23] Sears CL. A dynamic partnership: Celebrating our gut flora. Anaerobia. 2005;**11**:247-251

[24] Steinhoff U. Who Controls the Crowd? New Findings and Old Questions. 2005;**12**:14-18

[25] Kim E, Hong H, Choi K, Han Y, Kangwan N, Cho YC, Hahm KB. High concentrated probiotics improve inflammatory bowel diseases better than commercial concentration of probiotics. Journal of Food and Drug Analysis. 2012;**20**:292-295

[26] O'Hara AM, Shanahan F. The gut flora as a forgotten organ. EMBO Reports. 2006;**7**: 688-693

[27] Jewell AP. Is the liver an important site for the development of immune tolerance to tumours? Medical Hypothesis. 2005;**64**:751-754

[28] Mantis NJ, Rol N, Corthésy B. Secretory IgA's complex roles in immunity and mucosal homeostasis in the gut. Mucous Immunology. 2011;**4**:603-611

[29] Segal L, Blaser M. A brave new world: The lung microbiota in an era of change. Annual American Thoracic Society. 2014;**11**:521-527

[30] Conterno L, Fava F, Viola R, Tuohy K. Obesity and the gut microbiota: Does up-regulating colonic fermentation protect against obesity and metabolic disease. Genetics and Nutrition. 2011;**3**:241-260

[31] Al-Dohail MA, Hashim R, Aliyu-Paiko M. Evaluating the use of *Lactobacillus acidophilus* as a biocontrol agent against common pathogenic bacteria and the effects on the haematology parameters and histopathology in African catfish (*Clarias gariepinus*, Burchell 1822) fingerling. Aquaculture Research. 2011;**42**:196-209

[32] Son VM, Chang CC, Wu MC, Guu YK, Chiu CH, Cheng W. Dietary administration of probiotic, *Lactobacillus plantarum*, enhanced the growth, innate immune responses and disease resistance of grouper *Epinephelus coiodides*. Fish and Shellfish Immunology. 2009;**26**:691-698

[33] Faramarzi M, Kiaalvandi S, Lashkarbolooki M, Iranshahi F. The investigation of *Lactobacillus acidophilus* as probiotics on growth performance and disease resistance of rainbow trout (*Oncorhynkius mykiss*). American-European Journal of Science Research. 2011; **6**:32-38